Fire Safety Aspects
of
Polymeric Materials

VOLUME 8
LAND TRANSPORTATION VEHICLES

Report of

The Committee on Fire Safety
Aspects of Polymeric Materials

NATIONAL MATERIALS ADVISORY BOARD
Commission on Sociotechnical Systems
National Research Council

Publication NMAB 318-8
National Academy of Sciences
Washington, D.C.
1979

VOLUME 8 — LAND TRANSPORTATION VEHICLES

Fire Safety Aspects of Polymeric Materials

265 Post Road West, Westport, CT 06880

Printed in U.S.A.
Library or Congress Card No. 77-79218
ISBN 0-87762-229-9

NOTICE

The project that is the subject of this report was approved by the Governing Board of the National Research Council, whose members are drawn from the Councils of the National Academy of Sciences, the National Academy of Engineering, and the Institute of Medicine. The members of the committee responsible for the report were chosen for their special competences and with regard for appropriate balance.

This report has been reviewed by a group other than the authors according to procedures approved by a Report Review Committee consisting of members of the National Academy of Sciences, the National Academy of Engineering, and the Institute of Medicine.

This study by the National Materials Advisory Board was conducted under Contract No. 4-35856 with the National Bureau of Standards.

Printed in the United States of America.

FOREWORD

This volume is one of a series of reports on the fire safety aspects of polymeric materials. This work reported here represents the results of the first in-depth study of this important subject. The investigation was carried out by a committee of distinguished polymer and fire technology scholars appointed by the National Academy of Sciences and operating under the aegis of the National Materials Advisory Board, a unit of the Commission on Sociotechnical Systems of the National Research Council.

Polymers are a large class of materials, most new members of which are man-made. While their versatility is demonstrated daily by their rapidly burgeoning use, there is still much that is not known or not widely understood about their properties. In particular, the burning characteristics of polymers are only now being fully appreciated and the present study is a landmark in the understanding of the fire safety of these ubiquitous materials.

In the first volumes of this series the committee has identified the limits of man's knowledge of the combustibility of the growing number of polymeric materials used commercially, the nature of the by-products of that combustion, and how fire behavior in these systems may be measured and predicted. The later volumes deal with the specific applications of polymeric materials, and in all cases the committee has put forth useful recommendations as to the direction for future actions to make the use of these materials safer for society.

Harvey Brooks, Chairman
Commission on Sociotechnical Systems

ABSTRACT

This is the eighth volume in a series. The fire safety aspects of polymers are examined with primary emphasis on human survival. This volume is concerned with the polymeric materials used in subway, surface, and elevated urban railway vehicles; railroad vehicles; other rail vehicles (including unattended and semicontrolled type); buses, trucks; passenger automobiles; and miscellaneous vehicle types (including motorcycles and snowmobiles). Other volumes in the series deal with materials (state of the art); test methods, specifications, and standards; smoke and toxicity; fire dynamics and scenarios; aircraft (civil and military); buildings; ships, and mines and bunkers. A volume on elements of polymer fire safety and guide to the designer has been added to the series to pull together the disciplinary material of the first four volumes.

This report examines the fire safety aspects of those polymeric materials currently used, or expected to be used, in land vehicles that transport people and materials. Excluded from consideration are specific materials such as fuels, engine lubricants and other engine polymers and hydraulic fluids.

VOLUMES OF THIS SERIES

Volume 1	Materials: State of the Art
Volume 2	Test Methods, Specifications, and Standards
Volume 3	Smoke and Toxicity (Combustion Toxicology of Polymers)
Volume 4	Fire Dynamics and Scenarios
Volume 5	Elements of Polymer Fire Safety and Guide to the Designer
Volume 6	Aircraft: Civil and Military
Volume 7	Buildings
Volume 8	Land Transportation Vehicles
Volume 9	Ships
Volume 10	Mines and Bunkers

PREFACE

The National Materials Advisory Board (NMAB) of the Commission on Sociotechnical Systems, National Research Council, was asked by the Department of Defense Office of Research and Engineering and the National Aeronautics and Space Administration to "initiate a broad survey of fire-suppressant polymeric materials for use in aeronautical and space vehicles, to identify needs and opportunities, assess the state of the art in fire retardant polymers (including available materials, products, costs, data requirements, methods of test and toxicity problems), and describe a comprehensive program of research and development needed to update the technology and accelerate application where advantages will accrue in performance and economy."

In accordance with its usual practice, the NMAB convened representatives of the requesting agencies and other agencies known to be working in the field to determine how, in the national interest, the project might best be undertaken. It was quickly learned that wide duplication of interest exists. At the request of other agencies, sponsorship was made available to all government departments and agencies with an interest in fire safety. Concurrently, the scope of the project was broadened to take account of the needs enunciated by the new sponsors as well as those of the original sponsors.

In addition to the Department of Defense and the National Aeronautics and Space Administration, the total list of sponsors of this study now comprises Department of Agriculture, Department of Commerce (National Bureau of Standards), Department of Interior (Division of Mining Research, Health, and Safety, Bureau of Mines), Department of Housing and Urban Development, Department of Health, Education and Welfare (National Institute of Occupational Safety and Health), Department of Transportation (Federal Aviation Administration, Coast Guard), Department of Energy, Consumer Product Safety Commission, Environmental Protection Agency, and Postal Service.

The committee was originally constituted on November 30, 1972. The membership was expanded to its present status on July 26, 1973. The new scope was established after presentation of reports by liaison representatives which covered needs, views of problem areas, current activities, future plans, and relevant resource materials. Tutorial presentations were made at meetings held in the Academy and during site visits, when the committee or its panel met with experts and organizations concerned with fire safety aspects of polymeric materials. These site visits (upwards of a dozen) were an important feature of the committee's search for authentic information. Additional inputs of foreign fire technology were supplied by the U.S. Army Foreign Science and Technology Center and NMAB staff.

This study in its various aspects is addressed to those who formulate policy and allocate resources. A sufficient data base and bibliography has been supplied to indicate the breadth of this study.

ACKNOWLEDGEMENTS

A panel consisting of Dr. R.R. Hindersinn, Dr. G.R. Thomas, Daniel Pratt, Irving Litant, and Ms. Mauree Ayton, members and government liaison representatives of the National Materials Advisory Board Committee on Fire Safety Aspects of Polymeric Materials drafted in this report which was reviewed and finalized by the entire committee. Coordination was performed by Rear Admiral W.C. Hushing, USN (Ret.). The conclusions and recommendations are the sole responsibility of the committee.

Thanks are due to four manufacturers of land transport vehicles: Chrysler Corporation for detailed discussions with senior engineers involved in use of polymeric materials in automobiles; General Electric Company, Wayne Corporation, for discussions and a tour of school bus manufacturing facilities; Rohr Corporation, Flexible Division, for discussions and observations of the manufacturing process of general purpose buses.

Special thanks are due to Dr. George Thomas and Ms. Betty Sterling, Army Materials and Mechanics Research Center, for outstanding administrative support; and Irving Litant, Department of Transportation, for technical data, other information and access arrangements to government and industry activities. Presentations made by representatives of the Department of Transportation were helpful, particularly during the early stages of the committee's deliberations.

NATIONAL MATERIALS ADVISORY BOARD

Dr. William D. Manly
Senior Vice President
Cabot Corporation
Boston, Massachusetts

Dr. James W. Mar
Professor, Aeronautics and Astronautics
Massachusetts Institute of Technology
Cambridge, MA

Dr. Frederick T. Moore
Industrial Advisor
Industrial Development & Finance Dept.
World Bank, Washington, D.C.

Dr. Nathan E. Promisel
Consultant
Silver Spring, Maryland

Dr. Allen S. Russell
Vice President-Science and Technology
Aluminum Company of America
Pittsburgh, PA

Dr. Jason M. Salsbury
Director, Chemical Research Division
American Cyanamid Company
Wayne, New Jersey

Dr. John J. Schanz, Jr.
Assistant Director
Center for Policy Research
Resources for the Future
Washington, D.C.

Dr. Arnold J. Silverman
Professor, Department of Geology
University of Montana
Missoula MT

Dr. William M. Spurgeon
Director, Manufacturing and Quality Control
Bendix Corporation
Southfield, Michigan

Dr. Morris A. Steinberg
Director, Technology Applications
Lockheed Aircraft Corporation
Burbank, California

Dr. Roger A. Strehlow
Professor, Aeronautical and Astronautical Engineering
University of Illinois at Urbana
Urbana, IL

Dr. John E. Tilton
Professor, Department of Mineral Economics
Pensylvania State University
University Park, Maryland

Bureau of Standards, Washington, D.C.

NMAB Staff:

W.R. Prindle, Executive Director

Robert V. Hemm, Executive Secretary

NATIONAL MATERIALS ADVISORY BOARD
COMMITTEE ON FIRE SAFETY ASPECTS OF POLYMERIC MATERIALS

Chairman

HERMAN F. MARK, Emeritus Dean, Polytechnic Institute of New York, Brooklyn.

Members

ERIC BAER, Head, Division of Macromolecular Science, Case-Western Reserve University, Cleveland, Ohio

ALFRED E. BROWN, Celanese Corporation, New York.

JOHN M. BUTLER, Manager, Organic Polymer Synthesis, Monsanto Research Corp., Dayton, Ohio.

RAYMOND FRIEDMAN, Vice President and Scientific Director, Factory Mutual Research Corp., Norwood, Massachusetts.

ALFRED R. GILBERT, Manager, Organic and Polymer Chemistry Branch, Chemistry Laboratory, Corporate R&D, General Electric Company Schenectady, New York.

CARLOS J. HILADO, Research Professor, Institute of Chemical Biology, University of San Francisco, California.

RAYMOND R. HINDERSINN, Section Manager, Research Center, Hooker Chemical Co., Niagara Falls, New York.

WILLIAM C. HUSHING, Vice President, Kensington Management Consultants, Bath, Maine.

FRANK E. KARASZ, Professor and Co-Director, Materials Research Laboratory, Polymer Science & Engineering Dept., University of Massachusetts, Amherst.

TED A. LOOMIS, Professor & Chairman, Department of Pharmacology, School of Medicine, University of Washington, Seattle.

RICHARD S. MAGEE, Associate Professor, Department of Mechanical Engineering, Stevens Institute of Technology, Hoboken, New Jersey.

ELI M. PEARCE, Head, Department of Chemistry, Professor of Polymer Chemistry and Chemical Engineering, Polytechnic Institute of New York, Brooklyn, New York.

CHARLES F. REINHARDT, Director, Haskell Laboratory for Toxicology and Industrial Medicine, E. I. duPont Nemours and Co., Newark, Delaware.

ARNOLD ROSENTHAL, Celanese Research Company, Summit, New Jersey.

DONALD G. SMILLIE, Chief Structures Development Engineer, McDonnell Douglas Corp., Long Beach, California.

GIULIANA C. TESORO, Adjunct Professor, Massachusetts Institute of Technology, 278 Clinton Avenue, Dobbs Ferry, New York.

NICHOLAS W. TSCHOEGL, Professor Department of Chemical Engineering, Division of Chemistry and Chemical Engineering, California Institute of Technology, Pasadena California.

ROBERT B. WILLIAMSON, Associate Professor of Engineering Science, Dept. of Civil Engineering, University of California, Berkeley, California.

Liaison Representatives

JEROME PERSH, Staff Specialist for Materials and Structures, ODUSDRE(ET), Department of Defense, Washington, D.C.

ERNEST B. ALTERKRUSE, Chief, Department of Clinics, Moncrief Army Hospital, Fort Jackson, Columbia, South Carolina.

ALLAN J. McQUADE, U.S. Army Natick Laboratories, Natick, Massachusetts.

GEORGE R. THOMAS, Department of the Army, Army Materials and Mechanics Research Center, Watertown, Massachusetts.

HERBERT C. LAMB, Naval Facilities Engineering Command, Alexandria, Virginia.

DANIEL PRATT, Naval Ships Engineering Center, Prince George's Center, Hyattsville, Maryland.

GEORGE SORKIN, Naval Sea Systems Command, Washington, D.C.

JACK ROSS, Wright-Patterson Air Force Base, Ohio

BERNARD ACHHAMMER, National Aeronautics and Space Administration, NASA Headquarters, Advanced Research and Technology Division, Washington, D.C.

JOHN A. PARKER, National Aeronautics and Space Administration, Ames Research Center, Moffett Field, California

ARNOLD WEINTRAUB, Energy Research and Development Administration, Washington, D.C.

DAVID FORSHEY, Bureau of Mines, Washington, D.C.

CLAYTON HUGGETT, Center for Fire Research, National Bureau of Standards, Washington, D.C.

LESLIE H. BREDEN, Fire Safety Engineering Division, Center for Fire Research, National Bureau of Standards, Washington, D.C.

IRVING LITANT, Department of Transportation (DoT), Systems Center, Kendall Square, Cambridge, Massachusetts.

GEORGE BATES, JR., Department of Transportation, Federal Aviation Administration, Washington, D. C.

RALPH RUSSELL, Aircraft Division, DoT, Federal Aviation Administration, National Aviation Facilities Experimental Station, Atlantic City, New Jersey.

PAUL W. SMITH, Aviation Toxicology Laboratory, Civil Aeromedical Institute, Oklahoma City, Oklahoma.

THOMAS G. HOREFF, DoT Federal Aviation Administration, Washington, D.C. (Until September 1974).

DANIEL F. SHEEHAN, DoT, U. S. Coast Guard, Washington, D. C.

WILLIAM J. WERNER, Department of Housing and Urban Development, Washington, D. C.

DONALD L. MOORE, Department of Housing and Urban Development, Washington, D. C.

IRVING GRUNTFEST, Environmental Protection Agency, Washington, D. C.

NELSON GETCHELL, U. S. Dept. of Agriculture, Beltsville, Maryland.

WILSON A. REEVES, Cotton Finishes Laboratory, U. S. Dept. of Agriculture, New Orleans, Louisiana.

HERBERT W. EICHNER, Forest Service, U. S. Dept. of Agriculture, Forest Products Laboratory, Madison, Wisconsin.

RICHARD E. WIBERG, National Institute for Occupational Safety and Health, Rockville, Maryland.

JAMES RYAN, Consumer Product Safety Commission, Bethesda, Maryland.

JURGEN KRUSE, Materials Research Div., Office of Postal Science and Technology, Rockville, Maryland.

Technical Advisors

IRVING N. EINHORN, Flammability Research Center, Division of Materials Science and Engineering, Flammability Research Center, College of Engineering, University of Utah, Salt Lake City.

MAUREE W. AYTON, Science Information Analyst, Congressional Research Service, Library of Congress, Washington, D. C.

National Materials Advisory Board Staff

ROBERT S. SHANE, National Materials Advisory Board, National Academy of Sciences, Washington, D.C.

CONTENTS

CHAPTER 1

INTRODUCTION

1.1 Scope and Methodology of the Study

This volume is the eighth in a series of reports prepared by the Committee on Fire Safety Aspects of Polymeric Materials of the National Materials Advisory Board (NMAB).

The charge to the committee was set forth in presentations made by the various sponsoring agencies. Early in its deliberations, however, the committee concluded that its original charge required some modification and expansion if the crucial issues were to be fully examined and the needs of the sponsoring organizations filled. Accordingly, it was agreed that the committee would direct its attention to the behavior of polymeric materials in a fire situation with special emphasis on human-safety considerations. Excluded from consideration were firefighting, therapy after fire-caused injury, and mechanical aspects of design not related to fire safety.

The work of the committee includes: (1) a survey of the state of pertinent knowledge; (2) identification of gaps in that knowledge; (3) identification of work in progress; (4) evaluation of work as it relates to the identified gaps; (5) development of conclusions; (6) formulation of recommendations for action by appropriate public and private agencies; and (7) estimation, when appropriate, of the benefits that might accrue through implementation of the recommendations. Within this framework, functional areas were addressed as they relate to specific situations; end uses were considered when fire was a design consideration and the end uses are of concern to the sponsors of the study.

Attention was given to natural and synthetic polymeric materials primarily in terms of their composition, structure, relation to processing, and geometry (i.e., film, foam, fiber, etc.), but special aspects relating to their incorporation into an end-use component or structure also were included. Test methods, specifications, definitions, and standards that deal with the foregoing were considered. Regulations, however, were dealt with only in relation to end uses.

The products of combustion, including smoke and toxic substances, were considered in terms of their effects on human safety; morbidity and mortality were treated only as a function of the materials found among the products of combustion. The question of potential exposure to fire-retardant polymers, including skin contact, in situations not including the pyrolysis and combustion were addressed as deemed appropriate by the committee in relation to various end uses.

In an effort to clarify the understanding of the phenomena accompanying fire, consideration was given to the mechanics of mass and energy transfer (fire dynamics). The opportunity to develop one or more scenarios to guide thinking was provided; however, as noted above, firefighting was not considered. To assist those who might use natural or synthetic polymers in components or structures, consideration also was given to design principles and criteria.

In organizing its work, the committee concluded that its analysis of the fire safety of polymeric materials should consider the materials themselves, the fire dynamics situation, and the large societal systems affected. This decision led to the development of a reporting structure that provides for separate treatment of the technical-functional aspects of the problem and the aspects of product end use.

Accordingly, as the committee completes segments of its work, it plans to present its findings in the following five disciplinary and five end-use reports:

Volume 1 Materials: State of the Art
Volume 2 Test Methods, Specifications, and Standards
Volume 3 Smoke and Toxicity (Combustion Toxicology of Polymers)
Volume 4 Fire Dynamics and Scenarios
Volume 5 Elements of Polymer Fire Safety and Guide to the Designer
Volume 6 Aircraft: Civil and Military
Volume 7 Buildings
Volume 8 Land Transportation Vehicles
Volume 9 Ships
Volume 10 Mines and Bunkers

Some of the polymer applications and characteristics are in the classified literature, and the members of the committee with security clearances believed that this information could best be handled by special meetings and addendum reports to be prepared after the basic report volumes were completed. Thus, the bulk of the output of the committee would be freely available to the public. Considering the breadth of the fire safety problem, it is believed that exclusion of classified information at this time will not materially affect the committee's conclusions.

1.2 Scope and Limitations of This Report

This report examines the polymeric materials used in subway and elevated railway vehicles, surface vehicles (rail and urban), railway vehicles, other rail vehicles (including unattended and semicontrolled types), buses, trucks, passenger automobiles, weight handling vehicles, and miscellaneous vehicle types (including motorcycles and snowmobiles). For each case, the committee has attempted to determine:

1. What materials are used,
2. The parameters, physical and chemical, that influence flammability, smoke, and toxicity,
3. The material combinations, physical and chemical, that influence performance,
4. The use of materials in devices, subsystems, and systems,
5. The geometry, position, and environment of the material,
6. The contribution of the materials to system performance in normal and abnormal modes (fire).

Since much knowledge needed to make such determinations was lacking, the judgments of the committee are tentative and subject to revision; they represent only "best estimates possible" based on what is currently known and anticipated advances in the state of the art. (Inputs were delivered to the committee before July 1976, and literature references beyond that date generally are not included.) Additionally, it should be noted that while liaison representatives of sponsoring organizations attended the committee meetings bringing with them a wealth of data, background, and experience, the committee itself is solely responsible for the conclusions and recommendations presented in this report.

Although the relative priority of conclusions and recommendations was part of the committee's discussions, this report does not attempt to advise managers of resources on how to allocate them, vis-a-vis other demands on those resources.

Specific materials generally excluded from consideration in this report are fuels, engine lubricants and other engine polymers, and hydraulic fluids.

Recognizing the seriousness of the problem of fire safety of materials in all segments of society, the committee concluded that its work would be of value only if placed in the context of societal problems and their solutions. Accordingly, the committee assessed polymeric materials used in land transportation vehicles relative to:

1. Current materials knowledge and data,
2. Current test methods and standards,
3. Real world fire environments,
4. Status of knowledge smoke and toxicity,
5. Systems applications,
6. Potential for improvements,

The committee agreed on the nature of existing problems and deficiencies, but had some differences of opinion regarding the various solutions proposed and their priorities. It has nevertheless attempted to present a rounded picture of the present situation and what it believes to be the best current view in its conclusions and recommendations.

This report examines the fire safety aspects of those polymeric materials currently used, or expected to be used, in land vehicles that transport people and materials, i.e., automobiles, trucks, buses, railway cars, subway cars, surface urban transit vehicles, and certain new or proposed vehicles.

Because safety of life and limb of persons who use transportation systems was of prime concern during the committee's deliberations, its discussions, conclusions, and recommendations are heavily weighted in that direction. Another matter of deep concern was the safety of persons in the vicinity of a fire in a transportation vehicle, particularly system maintenance personnel, firefighters, rescue teams, and the general public in the vicinity of such fires or their atmospheric discharge (e.g., subway exhaust vents). Thus, although firefighting and rescue, system operation and maintenance, and system operation fall-outs are not within the province of the committee, those aspects most heavily involved as a result of polymeric material fires were subject to comment.

1.3 Committee Viewpoints

Since members of the committee are involved with materials research and development, applications, system design and evaluation and the liaison representatives deal with research and development, regulation, procurement, operations and analysis, the pertinent aspects of each material considered were subjected to a full spectrum of expertise. Moreover, full and extensive communications over the lengthy period of the committee's operation has provided an unusual data base to augment the members' expertise.

Many statements about the fire safety aspects of polymeric materials appear in each of the reports published as a result of the committee's study. Members of the committee wish to emphasize that such statements, including judgmental ones in regard to fire safety aspects of materials, especially end uses, apply only to the specific situations that pertain (e.g., suitability of a material from a fire safety point of view depends on many factors, including ease of access, ease of occupant, egress, proximity of ignition hazard, proximity of other materials, thermal flux and duration of ignition source, ambient oxygen partial pressure, and fire and smoke detection and suppression systems in place.)[1]

Although pertinent new knowledge and research results continually become available, it is anticipated that such new information will not change the overall assessment in the near term, but it may change specific statements or facts. Thus, undue emphasis should not be placed on any one specific statement regarding a material or experimental result and no statement should be taken out of context and applied to the use of identical materials in other situations. This viewpoint must be emphasized so that information that appears in all published reports of this committee's study is not misused by taking it out of context.

The sharply escalating costs of vehicles such as subway and railway cars and buses dictate that a substantial effort be mounted to reduce property damage and minimize costs of rehabilitation or restoration of these vehicles. Optimum choice of polymeric materials could make a major difference in this connection, and improved vehicle design, using all types of materials more effectively, would contribute significantly to this goal.

Service life of vehicles was another consideration that appears to have been subjugated to aesthetics, comfort, etc.; yet, in many cases, improved fire safety and increased service life are highly compatible.

This volume was written primarily for use by engineers and designers, manufacturers of transportation vehicles, manufacturers and developers of materials, transit authorities, local fire authorities, as well as government transportation administrators. It should be employed, however, as a supplement to existing data and rules. It represents the combined opinion of qualified experts in the field of polymeric materials and their

---

[1]This list is not all-inclusive, but only indicative of the kinds of concerns that must be considered in making a materials selection.

use in potential fire environments, but care must be taken not to take statements out of context and to recognize that all data, discussion, and analysis are, at best, current state of the art.

Volume 8 was written because the committee felt that better selection of polymeric materials and improved design for transport vehicles could:

1. Reduce substantially the fatalities from fires (transport vehicles now account for one third of all fire fatalities),
2. Reduce the tremendous financial losses resulting from fire in transport vehicles, whose unit costs, particularly in mass transit vehicles, have risen sharply,
3. Provide safe, more reliable transportation at less cost.

Natural polymeric materials such as wood, cotton and wool have been in use for a long time and a large body of scientific and empirical data concerning their properties and use has been accumulated. Despite such usage and many tests, the performances of natural polymers in dynamic fire environments are, at best, qualitatively described and are far less definitive than other performance characteristics such as tensile strength, elongation and molecular structure. Nonetheless, man has learned, largely by trial and error from fire disasters, to plan for natural polymer usages and cope with their fire characteristics.

Man-made polymers are of relatively recent development and are rapidly proliferating in usage. They offer advantages over natural polymers in many situations. In some cases their properties are well known and in others are being determined. Their fire related characteristics, however, are less well defined than those of the natural polymers, in some cases with significantly more serious potential consequences. Society has not had long years of experience or obtained the empirical data that provide the basis of planning and coping with the consequences of man-made polymer use, particularly in a fire environment. The continued rapid development of man-made polymers intensifies the severity of fire safety problems.

The deficiencies of specific man-made polymers (see Volume 1), inadequate fire test methods (see Volume 2), lack of sufficient knowledge of fire dynamics (see Volume 4) and lack of knowledge of the smoke and toxic effects (see Volume 3) are not well known to design engineers, vehicle manufacturers, transit authorities, and local fire authorities. Under such conditions, decisions to use or not to use various man-made polymers are based on considerations other than fire performance.

Man-made polymers have many advantages, including: optimization of desired properties, lower costs, reduction in weight (over metals), reduced production energy requirements, reduction in forming and manufacturing problems, as well as lower maintenance costs. These advantages, together with an apparent desire to emphasize aesthetics and comfort, have led to marked increases in the kinds and amounts of man-made polymers in virtually every transportation vehicle. The sharply increased fire load, with relatively unknown fire characteristics, poses a substantial problem that has not yet been fully realized, much less evaluated.

Fully (and partially) automated transport systems, such as new subways, and personal transport vehicles, lacking human sensors (operators), and severity to the problem. Such matters must be placed in proper perspective if the fire safety aspects of polymeric materials used in transportation vehicles are to be understood and evaluated.

1.4 Historical Perspective

Until the invention of the steam locomotive, ground transportation had not advanced very much for hundreds of years. The only means of travel on land, other than on foot or horseback, were the wagon, coach, and sled. The invention of the steam locomotive was followed by a tremendous surge in public mobility. New regions were opened for settlement because of the accessibility of products to markets. People achieved a new-found mobility that, while not very luxurious at first, was infinitely better and faster than bumping slowly over rutted roads in poorly suspended coaches. Freight moved in much greater quantities.

The accidents that occurred during the early years of railroading took many lives. Cars derailed, bridges collapsed, and fires occurred because of the spilling of hot embers from the pot-bellied stoves used to heat the shivering passengers in winter. The roadbed was not substantial, and the cars were flimsy and made of wood. Railroads became so popular and competition so fierce that the railroads soon began to look to the comfort of their passengers, if not to their safety. Passenger cars, well into this century, became opulent in the display of comfort not seen again after World War II, until the recent resurgence of passenger traffic under AMTRAK.

The streetcar made its appearance in the mid-19th century as a rail vehicle pulled by horses. When a useful steam engine became available, propulsion changed from horses to steam in some locales, but this venture was short-lived because the engines frightened horses drawing other vehicles that crowded the city streets. Besides, the electric motor came along in the 1880s, and the electric streetcar was born. Two modes of current pickup were devised — a third rail power source accessed through a slot in the street and the familiar overhead trolley wire. The latter became the predominant method of obtaining power.

The trolley streetcar was soon seen in most cities and even rivalled the railroad in intercity travel. For example, one could travel from Boston to Chicago by streetcar, making various changes from line to line. Speeds of over 80 mph were not uncommon. In a few cities, streetcars used an extensive tunnel system in the downtown area to avoid congestion of street traffic. Streetcars began to use subways in Boston in 1897, and in 1912, the first rail rapid transit train using a tunnel also was inaugurated, marking the beginning of a new means of mass transportation that was eventually followed in many of the large cities.

New subway systems now are being constructed or are in a stage of development in Washington, D.C., Baltimore, Atlanta, Dade County (Miami), and elsewhere. New streetcars called "light rail vehicles" (LRV) have been introduced in Boston, San Francisco, and Toronto in an attempt to improve urban mass transportation. The fire hazards associated with this mode of transportation will be addressed later in this volume, but it should be noted here that the fire hazard potential in LRV's is probably greater than for any other mode of ground mass transportation. (personal automobiles are excluded).

Changes caused by the advent of the internal combustion engine and the automobile, bus, and truck at the beginning of the century is well known. Improvement in roads and the freedom afforded by the automobile spelled the deathknell of the streetcar in all but a few localities. After World War II, railroad passenger service also was severely affected to the point of almost complete disappearance; a revival is now under way under the auspices of AMTRAK. For the most part, buses replaced the streetcars, but in fewer numbers. The automobile had become the predominant means of public transportation. Some buses remained tied to trolley lines (the trolley buses), but most are now diesel powered.

Spurred on by the fuel shortage and by environmental pollution problems, mass transportation advocates are seeking to make public transportation more attractive to commuters and long-distance travelers. AMTRAK is refurbishing railroad passenger service, and the Urban Mass Transportation Administration (UMTA) is providing billions of dollars for urban rapid transit and bus systems. The automobile industry is undergoing a massive change in the new emphasis on the smaller, more economical, fuel-efficient, non-polluting vehicles.

Now under development are innovations such as the air cushion and magnetically levitated vehicles, linear induction motors, a resurgence of the flywheel, and a host of new automobile engine designs. There also is greater use of automation. The next few years will see these and other developments applied in transportation vehicles. However, this report will not attempt to deal with fire hazards in projected developments where the use of polymeric materials cannot be foreseen clearly.

CHAPTER 2

CONCLUSIONS AND RECOMMENDATIONS

2.1 Introduction

Modern technology has permitted development of large and sophisticated land transportation units having elaborate interiors, components, and load bearing structures made from new polymeric materials. Sufficient knowledge about the fire safety aspects of these materials is lacking and, as a result, the danger to life and capital investment is high. Thus, development of a *sophisticated,* but practical, approach to fire prevention and control has become a necessity.

In each case where a potentially hazardous polymeric material is identified, a decision must be made: Should it be replaced with a more fire safe but usually more expensive or less satisfactory material? Should the hazard be minimized by design modification and user education? Should the hazard be accepted as an interim measure until a fully satisfactory substitute is developed? To make such a complex decision wisely, intimate knowledge of fire behavior is needed. This knowledge can be provided only through better understanding of fire dynamics and greater use of appropriate fire scenarios.

2.2 General Conclusions and Recommendations Applicable to All Vehicles

Conclusion: The use of polymers in land transport vehicles is rapidly expanding and has proceeded without adequate concern or consideration for fire safety. *Recommendation:* Consider vehicle fire safety from a systems basis to determine proper functional and safety requirements.

Conclusion: Data regarding polymer fire performance exist but these data are not being used effectively in the public interest. *Recommendation:* Analyze vehicle components from a fire safety standpoint, including consideration of operating conditions, functional requirements, and risk-reward values, to develop adequate specifications.

Conclusion: Vehicle design frequently reflects more concern for comfort and aesthetics than for occupant fire safety. *Recommendation:* Replace currently used polymers that are unsatisfactory in fire performance with materials more capable of meeting real-world fire threats.

Conclusion: The fire safety performance of passenger vehicles, particularly mass transportation units, needs significant improvement. *Recommendation:* Mount a major materials selection research program to develop fully functionally adequate and fire safe polymeric materials for most anticipated applications, after defining the fire goals of this program in advance, and provide for the use of adequate test methods to measure performance.

2.3 Conclusions and Recommendations, Fire Dynamics and Fire Scenarios

Conclusion: Despite serious deficiencies, current knowledge of fire dynamics is sufficiently advanced, if used in depth, to assist in improving vehicle fire safety, developing new and more useful fire tests, and predicting vehicle fire performance. To date, fire dynamics knowledge seldom has been used to assess the performance of land transport vehicles and associated systems. *Recommendation:* Increase use of current fire dynamics knowledge in mass transit vehicle design to prevent hazards and reduce losses.

Conclusion: The development and analysis of a fire scenario leads to the identification of critical stages in fire development, suggests opportunities for fire prevention, and directs attention towards various methods for fire control. *Recommendation:* Develop a wide spectrum of vehicle fire scenarios and quantify specific fire dynamics elements in these scenarios (e.g., fire spread and heat release rates). *Recommendation:* Prepare vehicle scenarios to permit generalization from the particular incident described and to provide the basis for exploration of alternative paths of the fire initiation and growth and for the analysis of the effect on fire performance of changes in material, design, and operating procedures. *Recommendation:* Train vehicle design engineers and transportation system fire safety personnel in the development and use of fire scenarios to enable them to more readily identify critical fire hazard elements and to determine appropriate safety measures.

2.4 Conclusions and Recommendations, Materials

Conclusion: Given sufficient oxygen and thermal energy input, all organic polymers will burn. Billions of pounds of synthetic and natural polymers are used annually in the United States without presenting unusual fire safety problems; however, some uses of polymeric materials in transport vehicles have seriously augmented the fire hazard. Many synthetic organic polymers burn in a manner different from that of the more familiar natural polymers such as wood, paper, cotton, or wool. Some synthetics burn much faster, some give off much more smoke, some evolve different noxious and toxic gases, and some melt and drip. Others burn less readily than the natural polymers. *Recommendation:* Initiate a program to define the critical overall fire safety parameters (flammability, smoke and toxic gases, etc.) of polymer based materials. *Recommendation:* Support approaches to improve the fire safety of the high volume and low cost polymers. *Recommendation:* Initiate programs to determine the relationship of chemical and physical components of polymeric materials to the evolution of smoke and toxic gas formation. *Recommendation:* Develop a sound education program for all age levels to better acquaint the public with the fire safety aspects of polymeric materials and products.

Conclusion: The flammability of many polymers has been improved by the incorporation of hydrated alumina and/or compounds containing halogens, phosphorus, and/or antimony. *Recommendation:* Develop an assessment technique to quantify the potential threats of additives, monomers, polymers and other materials used in the synthesis or modification of polymers to improve their fire safety characteristics.

Conclusion: Many polymeric materials with improved flammability characteristics are deficient in other ways (e.g., costs are high, fabrication is difficult, and combustion products are toxic and corrosive). *Recommendation:* Increase the development effort on char forming systems with particular emphasis on lowering the fabrication cost.

Conclusion: Intumescent coatings can be used to enhance the fire safety of some polymeric products. *Recommendation:* Expand materials and application studies on intumescent coatings with emphasis on lowering cost and improving performance.

Conclusion: Designers and engineers of land transport vehicles, like many other designers, architects and engineers, generally lack training in and knowledge of the fire performance properties of polymeric materials and the fire safety problems associated with them. *Recommendation:* Require greatly increased emphasis on fire safety of polymeric materials in the scholastic curriculum for designers and engineers and maintain this emphasis through on-the-job training and review.

Conclusion: Some synthetics, burning in a manner different from that of the more familiar natural polymers, often present unexpected complications in a fire situation with which the average person cannot cope; this is particularly a problem in unusual circumstances such as in confined spaces. *Recommendation:* Develop fire safety systems analysis methods, including the use of scenarios, to guide materials selection and to permit the overall fire safety assessment to be based on material design, environment, detection, and fire control factors. *Recommendation:* Initiate programs to increase basic knowledge of the relationship between the chemical and physical properties of polymers and fire dynamics parameters and the way this relationship is affected by aging.

Conclusion: Elastomeric polyurethane cushioning is widely used in transport vehicles and burns readily even when fire retardants are incorporated in the resin; new materials and/or new approaches using current material are urgently needed. Carpet systems vary greatly in their response to a fire, and systems that are safe for uses in urban transit cars, buses, etc., need to be developed or defined. The use of carpeting on walls or ceilings of subway cars, buses, etc., is particularly undesirable. *Recommendation:* Implement programs (chemical, materials, or engineering) to improve the fire safety characteristics of cushioning systems including consideration of smoke, toxic gases, flammability, and manufacturing and operational practicality.

Conclusion: Blends of cotton and polyester fibers make very desirable and economical fabrics for seat covers, etc.; however, such blends cannot be fire retarded with effective and economical treatments. *Recommendation:* Establish incentives to accelerate the introduction and commercialization of new materials with improved fire safety characteristics.

Conclusion: Concern exists about potential fire hazards associated with the rapidly increasing use of polymeric structural and insulating foams in such vehicles as urban transit cars. *Recommendation:* Create an overall program to categorize and communicate the goals and results of government-supported work on the fire safety of polymeric materials.

2.5 Conclusions and Recommendations, Tests Methods, Specifications and Standards

Conclusion: Test methods available to regulatory agencies are inadequate to provide guidance for selection of polymeric materials to be used in vehicles. *Recommendation:* Contribute to the improvement of fire safety of vehicles and transit systems by developing the data needed to improve test requirements, test specifications, test data, and test extrapolation methods.

Conclusion: The flammability and smoke emission of materials used in public transportation vehicles such as rapid transit and railroad passenger cars currently is covered only by recommended guideline specifications, not federal regulatory standards. *Recommendation:* Develop and rapidly implement regulations concerning allowable parameters for flammability, smoke emission, and toxicity.

Conclusion: Only one federal regulatory standard for ground transportation, Motor Vehicle Safety Standard (MVSS 302), applies to automobiles, trucks, buses, and recreational vehicles. This standard prescribes a test method that tests materials only in a horizontal orientation and is considered by test experts to be almost totally ineffective in providing fire safety in a real fire situation. *Recommendation:* Develop new standards that will better define the fire performance of combustible materials in vehicles (e.g., standards recognizing that materials oriented vertically may spread flame an order of magnitude faster than the same oriented horizontally).

2.6 Conclusions and Recommendations, Smoke and Toxicity

Conclusion: As the diversity and amount of polymeric materials used in vehicles increase, the problems presented by the generation of smoke and toxic gases in a fire also increase. Objective information defining the extent and nature of this hazard is not available. Although it is known that the thermal decomposition products from synthetic polymers contribute to the overall hazard during vehicle fires, the degree to which each product contributes is difficult to establish because the relationships of chemical and physical composition to smoke and toxic gas formation during combustion are not well understood. Carbon monoxide is a well established major toxic hazard in polymer fires, but current data suggest that, under both clinical and experimental conditions, thermal decomposition products other than carbon monoxide can be major contributors to the hazard to human survival. Species and quantities of pyrolysis and combustion products vary, in most cases, depending on the temperature to which materials are subjected and the amount of oxygen available. *Recommendation:* Increase in scope as well as effort and closely monitor, for application to land passenger vehicles, the research program established to develop criteria and practices for determining the degree to which polymers contribute to human morbidity and mortality. *Recommendation:* Expand research programs directed toward an assessment of the hazards from toxic products to more clearly and rapidly define the effects of varying temperatures and oxygen availabilities during combustion of polymeric materials. Smoke and toxic gases are often more dangerous to human health and survival than the thermal effects of combustion (burns). Some toxic gases are not discernible by normal human senses. *Recommendation:* Develop and implement an educational program to advise the public and emergency personnel of the increasing hazard that smoke and toxic gas pose to human health.

2.7 Conclusions and Recommendations, Fixed Guideway Vehicles

Conclusion: Fixed guideway vehicles operate under conditions that can rapidly develop into hazardous situations. Many of the materials currently used in ground mass transportation vehicles are potentially hazardous in that they reduce the effectiveness of fire protection in the transit systems and contribute to increased fire hazard from the production of smoke and toxic gases. *Recommendation:* Use those polymeric materials that, by testing and comparison, are judged to be the most fire retardant and have the lowest smoke and toxic gas emission rates; even these materials should be used sparingly, consistent with comfort and service ability. *Recommendation:* Develop a methodology to determine the overall fire threat and fire hazards of a given system and to provide the data that will be used during the design phase to mitigate the fire threat. *Conclusion:* In giving consideration to aesthetics and comfort in recent rail car designs and construction, there have been instances of improper and excessive use of certain polymers; this has resulted in unnecessary fire loads and increased fire susceptibility. *Recommendation:* Fire-harden and provide the air conditioning plenum in passenger vehicles with closures having fail-safe fusible links. *Recommendation:* Replace polyvinyl chloride formulations currently in wide use as wall ceiling panels with more flame retardant, less smoke-emitting plastic panels; consideration also should be given to the use of metal panels with plastic coatings to provide desired colors and texture.

Conclusion: Because mass transit vehicles operate in tunnels and other hazardous zones, and the materials of vehicle construction must be the most resistant to ignition, consistent with necessary tradeoffs of risk versus cost. *Recommendation:* Design vehicles and transit systems using materials with the greatest potential for minimizing the threat of fire.

Conclusion: The floor of a fixed guideway passenger vehicle is the prime fire barrier between the passenger compartment and the operating components beneath the car where most fires originate. *Recommendation:* Require polymeric materials used in floor construction to provide the same resistance to penetration by fire as current plymetal (metal-wood composite) construction.

Conclusion: The continued operation of a passenger vehicle electrical system under fire conditions is a most important factor in total fire safety. *Recommendation:* Implement the recommendations of the several studies based on the "smokeless cable" concept to provide protection from the fire hazards of electrical insulation and to insure continuity of electrical service during fire.

Conclusion: Vehicle designers, manufacturers, and operators have inadequate knowledge of the fire-related properties of the polymeric materials being used. *Recommendation:* Make available automatically operated fire extinguishment devices in those situations where trained personnel are not in attendance as well as other fire extinguishment resources at frequent intervals at the wayside.

Conclusion: Radiation of heat from the walls and ceiling of tunnels has been largely ignored in calculating the effects of heat loads on the fire safety of materials used in vehicle construction. *Recommendation:* Take into account the essentially closed system of a tunnel in making calculations of total heat loads imparted by a burning vehicle.

Conclusion: The use of presently known flexible polyurethane foam systems in seat cushions is not consistent with overall fire safety; polychloroprene (neoprene) foams currently are the only reasonable substitute cushion materials. *Recommendation:* Do not use polyurethane foam in seat cushions; instead use polychloroprene foams.

2.8 Conclusions and Recommendations, Buses

2.8.1 Matters for First Attention

Conclusion: Polyurethane foam seat cushions are a serious hazard in current buses. *Recommendation:* Prohibit the installation of polyurethane foam seat cushions in new buses, and refurbish current operational buses with better materials. Specify a better material, such as fire-retarded neoprene foam, until such time as a more suitable material becomes available.

Conclusion: Floors and other structures serving as fire barriers have some penetrations that are poorly designed or are constructed of unsatisfactory materials (from a fire safety standpoint). The most serious situation in this regard is the use of fiber reinforced polymer wheel well covers. *Recommendation:* Require that wheel well covers made of fiber-reinforced polymers provide fire resistance equivalent to that of steel to prevent wheel (tires and brakes) fires from penetrating the passenger compartment. Specifications for wheel well covers should match those of the structural floor regarding fire performance.

Conclusion: Electric wiring, particularly that which carries current to operating devices (and thus is susceptible to large circuit currents), run behind interior body paneling, is a potentially grave and strong ignition source; its proximity to polymeric insulations and interior paneling can add to the hazards. *Recommendation:* Run electric power wiring, insofar as practical, in metal conduits. When this is not possible, the routing of such wiring should be carefully designed and monitored in the manufacturing process to ensure that the wiring is protected from cuts and chafing and that it is properly fused; it should be protected from contact with polymers used in insulation and body paneling, and the insulation of the wiring should have good fire performance characteristics.

2.8.2 Items for Progressive Improvement of Fire Performance of Buses

Conclusion: The use of composite polymers having good fire performance offers many advantages, particularly in bus construction. The polymers currently used in bus bodies do not have good fire performance. *Recommendation:* Initiate a program to develop, qualify and approve polymers having good fire performance for bus construction. Replace the polymers to be used in new bus bodies with steel or fire retardant composites providing equivalent fire protection.

Conclusion: Experimental and prototype buses supported by the (Urban Mass Transportation Administration (UMTA) and the manufacturer have not been subjected to rigorous fire performance analysis using scenarios; limited scenario analysis leads to a tentative conclusion that materials selected and used in such experimental buses will not give adequate performance for fire safety. *Recommendation:* Without UMTA grants to bus operating authorities for new transit bus purchases (including school bus usage of such buses) until purchase specifications comply with UMTA safety standards. *Recommendation:* Require a rigorous fire-biased analysis, using scenario techniques, for each new bus design; withhold UMTA funds for development or acquisition until such analyses quantify the risks and show that they have been reduced to an acceptable level.

Conclusion: Many polymers are not being properly used in bus construction (e.g., those used in carpeting on side or end walls or on the ceiling). *Recommendations:* Analyze new and existing bus designs with the aim of eliminating improper or unnecessary use of flammable polymeric materials (e.g., the use of carpeting on vertical walls or in overhead positions).

2.9 Conclusions and Recommendations, Passenger Vehicles

Conclusion: The estimated annual $135 million property loss and about 500 fatalities directly associated with passenger car fires is sufficiently serious to warrant considerable effort to improve the fire safety characteristics of these vehicles. Passenger compartment furnishings are almost 100 percent polymeric materials (predominantly flexible polyurethane, polyvinyl chloride, polypropylene, and SBR elastomers). Polymeric materials are major contributors to the high frequency of passenger compartment fires. The increasing substitution of flammable polymers for metal in passenger car construction can be expected to increase the frequency and severity of passenger car fires unless adequate steps are taken to reduce the flammability of these materials by developing and enforcing stringent flammability test standards. The replacement of MVSS 302 with a more stringent flammability standard can be expected to significantly reduce the incidence and severity of passenger compartment fires at only moderate cost.

Recommendation: Carefully evaluate the flammability characteristics of passenger vehicles with bodies constructed entirely or largely of synthetic polymers on full-scale mock-ups before such vehicles are mass produced for general consumption. Require all polymeric materials used in automobile passenger compartments to pass FAR 25.853, sections a and b (sections b-1, b-2, and b-3 are not considered suitable), except for those components that can be shown to create no appreciable fire hazard. Undertake a study to determine the effect of plastic fuel tanks on overall fire safety.

2.10 Conclusions and Recommendations, Trucks and Special Purpose Vehicles

Conclusion: Use of plastics and reinforced plastics in highway vehicles such as trucks, recreational vehicles, and motorcycles is increasing and will continue to increase for a variety of reasons. For example, the increased fatigue life of polymers reduces the need for replacement parts; factors such as reduced corrosion lower maintenance costs, the lower density of plastics [compared to metals] permits reduction in the overall weight of the vehicle. This reduction in weight lowers fuel cost and consumption for the operator, resulting in energy saving for the nation. MVSS 302 was developed from statistics for vehicles constructed primarily out of metals. There is real doubt whether or not these statistics will remain constant during the ongoing shift from nonflammable metals to plastics. *Recommendation:* Reexamine MVSS for fire safety aspects in the control of the system currently are being used and in those in which they are intended to be used. Reexamine MVSS in light of the materials now in use and those expected to be used in transport systems in the future.

Conclusion: Fire scenarios coupled with qualitative fire dynamic considerations have not been applied in analyses of conditions as they will exist and have not been utilized in the prediction and prevention of future hazards. *Recommendation:* Verify fire scenarios through experiments and assign test methods and material property values on the basis of these experimental scenarios. For an unverified scenario, assign tentative test methods and material property values on the basis of a worst case situation.

Conclusion: A recurring problem appears to be that fuel storage facilities on vehicles contribute to ignition and the spread of a fire under impact of a collision. *Recommendation:* Examine the crash worthiness of all fuel systems involving polymeric materials with a view toward eliminating the fuel system as an ignition source, as an aid to flame spread, or as a source of fuel for the fire.

CHAPTER 3

FIRE DYNAMICS AND FIRE SCENARIOS

3.1 Introduction

This chapter examines two major, interrelated aspects of fire safety — fire dynamics and fire scenarios. Neither of these approaches is being used sufficiently to realize available potential benefits in increased fire safety for transport vehicles, particularly when such vehicles are part of a complex subway mass transit system. In each situation an approach to improved fire safety must be selected from a set of alternatives; to make the choice wisely, an intimate knowledge of fire and fire behavior is required. An understanding of fire dynamics, coupled with the use of appropriate fire scenarios, can provide such knowledge. A more detailed examination of the subject of this chapter is undertaken in Volume 4 of this report.

The term "fire dynamics" refers to the scientific description of fire phenomena in quantitative terms. Phenomena of interest include ignition, flame spread, maximum burning rate, growth, intensity, products of combustion, movement of combustion products through a vehicle (or vehicle right-of-way), detection, extinguishment and effect on humans. "Fire scenarios" are generalized descriptions of fire incidents that should include all details relative to prefire considerations (system design, materials choice, fire detection, fire suppression components, etc.), the fire itself, and subsequent behavior of people and protection devices. An infinite number of scenarios can be developed from a given fire situation; however, development of a few provide virtually all of the analytical information necessary to formulate useful plans for operations.

An alternative to the detailed scenario approach is the statistical approach: "Collisions, electrical faults, and vandalism cause x, y, and z percent respectively of all fatal fires in vehicles." (Such a statement could be considered to consist of three highly condensed scenarios.) While such statistics are useful, they fail to provide the needed level of information on which to base decisions to replace, to redesign, to educate, or to tolerate a given hazard. For example:

> Introduction of a fire retardant into a polymeric material may reduce the probability of igntion, but also increases the optical density of the smoke produced if ignition does occur. Under what conditions does the fire retardant result in an increase or decrease of overall fire safety? Material A is harder to ignite than material B but, once ignited, spreads a flame much more rapidly. Under what conditions is A more hazardous than B?

To analyze each case, one clearly needs to know the chemical nature of the combustible; the geometry of the vehicle compartment where the fire occurs and the adjacent compartments; ventilation factors; the location of occupants; the time lag until combustion products reach a human automatic detector; and the possibilities of extinguishment, escape, or rescue. These variables usually are not included in fire statistics.

The availability of statistics varies considerably among the various modes of ground transportation. Automobile accident statistics are well recorded both by the state authorities and insurance companies. Insurance authorities, however, have been very reluctant to disclose accident statistics and are not required by local or state governments to do so. On the other hand it is known that many small fires are extinguished by transit personnel, but nowhere can records of the actual kind, number, or severity of occurrences be found. Fire department records are of limited use because they generally detail only extinguishment procedures.

Reporting of fires by railroads also has been poor. The old so-called "T-sheets," which were accident reports required by the Federal Railroad Administration (FRA), generally did not list a railroad fire if that was not the prime reason for an accident. In 1975, the Federal Railroad Administration issued its "Guide for Preparing Accident/Incident Reports," which requires all rail systems to report all accidents in detail on a monthly basis, and in time, these reports may provide valuable information concerning sources of ignition, fire propagation, the extent of damage, and the relative difficulty of extinguishing fires.

Despite some serious deficiencies, the knowledge of fire dynamics is good enough at this time to make some useful contributions to fire safety. The complete analysis of a scenario, for example, should include what actually happened as well as alternative events that could easily have occurred under slightly different circumstances. Fire dynamics knowledge is the only basis for predicting alternative variations of a scenario, short of performing realistic full-scale fire tests. In addition, it may be desirable to perform a "partial full-scale" fire test in which, e.g., only one end of a subway car is equipped completely and the rest of the car is bare (to reduce testing cost), and an analysis of the fire dynamics involved should indicate the degree of validity of such a test. Obviously, the relation of laboratory test methods to realistic fire behavior involves fire dynamics knowledge, and even the current incomplete state of knowledge should be brought to bear on such tests, at least to identify qualitatively the key parameters involved (radiant flux level, ventilation, etc.). Preliminary evaluation of proposed design changes for fire-hardening also require fire dynamic inputs.

It should be noted that fire dynamics and fire scenario generation and analysis are really two different disciplines, each requiring specialists with different training and aptitude.

3.2 Fire Dynamics

3.2.1 What is Fire Dynamics?

In broad terms, fire dynamics involves quantitative descriptions of fire phenomena on a scientific basis. These phenomena may be approached on any one of four levels.

On the first level, fire is governed by:

1. The chemical thermodynamics and stoichiometry of combustion,
2. The chemical kinetics of pyrolysis and combustion reactions,
3. The transfer of combustion energy by conduction, convection, and radiation,
4. The motion of combustion gases as driven by bouyancy, thermal expansion, or mechanical force and modified by constraining walls, viscous effects, inertia, the nature of turbulence, the properties of hot gases, etc.

On the second level, fire may be broken down into a series of phases or stages such as:

1. Smouldering.
2. Spontaneous ignition.
3. Piloted ignition.
4. Horizontal or downward flame spread over solids.
5. Upward flame spread over solids.
6. Flame spread over a liquid below its flash point.
7. Flame spread over a liquid above its flash point.
8. Burning rate of a liquid pool.
9. Burning rate of a solid slab.
10. Formation of toxic species in a diffusion flame.
11. Formation of aerosols in a diffusion flame.
12. Radiation emitted by a diffusion flame.
13. Extinguishment by heat loss.
14. Extinguishment by reduction of oxygen.
15. Extinguishment by chemical inhibitors.

Research on these topics is largely, but not entirely, academic. Results would be of general interest to those outside of the fire community concerned with combustion (e.g., in connection with vehicle ventilation design).

On the third level, fire dynamics is concerned with complex processes generally involving the interaction of two or more of the subjects listed above. Some examples are:

1. Burning rate of an object as influenced by radioactive feedback from the environment.
2. Burning rate of an object in a noncombustible compartment as influenced by ventilation of the compartment.
3. Generation of incomplete combustion products as influenced by either of the above conditions.

4. Mutual interactions of two adjacent burning objects.
5. Spontaneous ignition of pyrolysis gases from a hot object as influenced by turbulent free convection and mixing.
6. Properties of smoke from a fire in a compartment as influenced by the mixing, cooling, and "aging" or agglomeration that occurs in the interval between generation and arrival of smoke at a detector station.
7. As cooling occurs absorption of toxic gases from fires by aerosols generated in the fire.
8. The effects of radiant emissions, transmission, and absorptions in a compartment at a pre-flashover stage of fire.
9. The effects of physical scale on fire turbulence, on fire radiation, and ultimately on fire behavior.
10. The relative effects of chemical kinetics and physical factors on a fire near an extinguishment condition.
11. Determination of the source of undesirable products of incomplete combustion in a fire (i.e., surviving initial pyrolysis products vs. products formed in gas-phase reactions in or near a flame).
12. Effects of long-term exposure of materials to various ambient conditions on subsequent fire behavior.
13. Identification of the mechanisms involved in the interaction of water spray with a fire (e.g., cooling of a pyrolyzing solid, cooling and/or diluting of a flame with steam, prewetting of adjacent fuel, absorbing radiation entering into the flame chemistry via $C+H_2$ or $CO+H_2O$, entraining air with the water spray, superheated droplets exploding in molten polymer and spattering fuel and thereby increasing combustion).

Successful fire dynamics studies of these procedures will have major value in strengthening the engineering judgment needed to validate test methods for realistic fire hazards. Academic researchers often tend to avoid this third level of problems because many of them feel there are still so many major unknowns in the second and even the first levels of problems that this third level is too difficult to treat properly.

Except for mentioning toxicity, fire has been discussed above as if it occurred in an uninhabited world. Thus, the fourth level of fire dynamics involves the interaction of fire phenomena with human response. Problems such as the following exist:

1. Actions of humans that lead to ignition.
2. Sensory detectability of a fire (including smell and sound).
3. Vision as affected by smoke or lachrymatory gases.
4. Panic or confused thinking as induced by fire phenomena.
5. Human ability to control fire at various stages of development as governed by training, equipment available, panic, etc.
6. Burn damage of skin by garments as influenced by rate of flame spread, melting, dripping, etc.

7. Toxicity including prefire condition of victim (blood alcohol content, heart or circulatory disease, etc.).
8. Toxicity including combined (synergistic) or sequential effects of various toxic species of combustion products.

It is recognized that study of human behavior may well be considered as outside the scope of a treatise on polymeric materials behavior, but it seems reasonable to call attention to this important aspect of the problem to complete the catalogue of what needs to be known.

It should be noted that in any event, scanty funds, personnel, and other resources are currently available for the study and development of fire dynamics. This results in slow progress toward adequate knowledge.

3.2.2 Critical Fire Dynamics Elements

3.2.2.1 Ignition

Ignition of a combustible material is the first step in any fire scenario and therefore is important to fire prevention. Furthermore, once a fire has started, the ignition delay times of other materials, coupled with flame spread rate, will affect the rate at which the fire develops.

Most polymeric materials can be made to ignite and burn if they are sufficiently heated in the presence of sufficient oxygen. Consequently, from the point of view of fire safety, it is desirable to know how long it will take a particular polymer to ignite under various fire conditions (e.g., oxygen concentration, environmental gas temperature, and heating rate).

3.2.2.2 Flame Spread

The propagation of a flame over a combustible solid is an extremely complex process. However, since the flame spread rate is readily measured, a large volume of experimental flame spread rate data has been obtained over the past decade. Nevertheless, uncertainty still exists as to which physical and chemical parameters exert dominant effects.

Since heat must travel ahead from the flame to the unignited material in order to propagate the flame, it is clear that certain heat transfer modes must be involved. However, the relative importance to flame spread rates of conduction or convection in the gas phase, conduction in the condensed phase, and radiation in the gas phase is not known even in the simplest cases. The motion of the gas at the leading edge of the flame is of potential importance. The gas phase chemistry, including oxidant concentration, inert dilutent, and pressure as well as interdiffusion of reactants, also must be considered. Finally, the influence of solid-phase surface chemistry, thermal properties, and geometry cannot be ignored.

The results from many experimental investigations have been summarized by Friedman (1968) and by Magee and McAlevy (1971). These surveys indicate that the flame spread rate is affected by many physical and chemical parameters. Among the more important of those are

1. Physical and geometrical parameters;
    a. Orientation of surface;
    b. Direction of propagation;
    c. Thickness of specimen;
    d. Surface roughness;
    e. Presence of sharp edges;
    f. Initial fuel temperature;
    g. Environmental pressure;
    h. Velocity of environment;
    i. External radiant flux;
    j. Humidity;
    k. Specimen size;
2. Chemical parameters
    a. Composition of solid;
    b. Composition of atmosphere;

Most studies, however, have been concerned with horizontal or downward propagation even though the very important upward spread process is several orders of magnitude faster and appears to be controlled by different processes.

3.2.3 Dynamic Analysis

The current state of the art provides only limited ability to predict the rate at which smoke and combustion products move from flaming or smoldering areas to more remote parts of transport vehicles or the rate of dilution as this movement occurs. These rates will differ in a moving vehicle from those in a stationary vehicle, and the draft in a tunnel will have a major effect. The importance of smoke transport rate to smoke detector location and on the escape potential for passengers should be obvious. Fortunately, some research now is under way to develop scientific tools for this type of prediction. Dynamic analysis of buoyant motion of smoke and fire gas is in its infancy; if this tool were developed, it could become capable of treating successively more complex geometries.

3.2.4 Extinguishment

One question concerning the fire hazard associated with a given polymeric material is: Once the material has ignited and begun to burn, how easily can the fire be extinguished? The answer, ease of suppression, will depend on the extinguishment approach employed as well as on the fire conditions prevailing at that time.

Various approaches to the problem of extinguishment of unwanted fires are:

1. Isolation of the polymer fuel.
2. Isolation of the oxygen.
3. Cooling the solid polymer fuel.
4. Cooling or diluting the gas phase.
5. Inhibition of the chemical reaction homogeneously.
6. Inhibition of the chemical reaction heterogeneously.

The extinguishment of fires by water has long been practiced. Since water usually is available, inexpensive and generally effective, many subway tunnels and public stations are protected by automatic sprinkler systems that employ water as the extinguishing agent. Water, being nontoxic and having a very high heat absorbing capacity, will continue to be employed as the suppressing agent for many fire situations.

Water, steam, and fogging nozzles are designed to achieve fire suppression by cooling the hot combustion gases. One study has reported that, under certain conditions, this approach is less efficient than cooling the condensed phase directly. Further study of minimum water requirements for both application modes is needed under various fire conditions.

Carbon dioxide extinguishers function primarily by diluting the oxidizer or by isolating the oxidizer from the fuel; however, benefit also may come from some cooling of the gas phase and condensed phase as well. Regulations frequently specify the size and number of CO_2 extinguishers that are required in various types of vehicles; yet data on the amount of extinguishing agent and the application rate necessary to suppress different types of fires are lacking.

Homogeneous chemical inhibitors (e.g., halogen-containing compounds such as CF_3Br [Halon 1301]) have been found effective in suppressing hydrocarbon fires as well as Class A fires as long as they are not deepseated. The effectiveness of these agents and the suppression mechanisms involved have been the subject of extensive study. The mechanisms of suppression are still open to debate and information on the dependence of inhibition requirements on various fire parameters, fuel type, fire intensity and other factors is lacking.

Compounds such as sodium and potassium bicarbonate are the principal heterogeneous chemical inhibitors. The effects of these inhibitors on fire suppression also have been reviewed. A recent study seems to have established that the inhibiting mechanism is not heterogeneous (occurring either on the surfaces of the alkali metal salt particles or on the surfaces of condensed oxide particles produced when the salt decomposes), but is homogeneous, possibly through chain breaking by alkali metal hydroxides. Further support of a homogeneous inhibition mechanism for potassium salts has been offered.

3.2.5 Application of Fire Dynamics to Test Method Development

In general, contemporary knowledge of fire dynamics is insufficient to permit identification of its use in test method development. Also, most current tests were developed either long ago when knowledge was even more primitive or at times when enormous pressure existed to develop test methods in a matter of months, permitting no time for a fire dynamics approach (i.e., tests often are introduced in response to a specific disaster rather than as part of a long-range plan).

One type of test method is the fire modeling approach in which a physical model of a fire situation is reduced in scale while maintaining geometric similarity and preserving the important chemical and thermodynamic properties of the materials. If such an approach is successful, the benefits to fire testing are obvious. Unfortunately, progress has been very slow.

The difficulties of modeling a fire by reducing scale arise in several ways: (1) very small fires are laminar while larger fires are turbulent; (2) as far as fluid mechanics is concerned, the ratio of buoyancy forces to viscous forces in the convective flow of fire gases is size-dependent; (3) the radiant emission and self-absorption of the flame are size-dependent; and (4) the gas-phase time scale in the fire is shorter for small than for large fires, with possible additional effects on the formation of intermediate combustion products.

These formidable difficulties have not fostered confidence in model test results; however, it may be hoped that valid modeling procedures can be developed for at least some aspects of fire behavior as understanding of pertinent fire dynamics improves. Further, any errors of distortions introduced by the modeling might be compensated for by varying other parameters such as ambient temperature, pressure, oxygen concentration, or ambient radiation. It has been suggested that the effect of gravitational force may be modeled by use of a centrifuge. Some progress has been made in pressure-modeling techniques.

3.2.6 Current State of Fire Dynamics Knowledge

Fire dynamics knowledge relating to land transport vehicles has grown very slowly. Very few full-scale vehicle fire tests have been conducted. Lesser tests prescribed by regulations and customer specifications have some application to dynamic fire situations and have some use in predicting fire and smoke risks imposed on passengers (and cargo). Although full-scale fire tests of vehicle interiors may be useful in screening out unsatisfactory components and polymers, they have limited application for developing satisfactory polymeric materials to be used in land mass transport vehicles. A substantially greater effort to develop fire dynamics knowledge applicable to land transport vehicles is clearly required. Most current knowledge has been gained by analysis of actual vehicle fires.

3.3 Fire Scenarios

3.3.1 Introduction

Availability of detailed fire scenarios is beneficial not only as an aid in analyzing specific accidents or in devising specific fire test methods, but also as a general guide in developing standards, codes, and regulations and in planning research programs. Scenarios have maximum utility if: (1) they represent accident modes that cause a significant fraction of the annual casualties or loss from fire, and (2) they provide sufficiently detailed information to permit useful analysis.

As to the frequency of occurrence of a given scenario, statistical analysis of accident data is helpful as a guide, but only very limited data are available and technological change may occur so rapidly that the time lag between the introduction of a new material, product, or structure and development of a statistically significant accident history may be unacceptable. In view of both these factors, judgment and extrapolation are very important in developing meaningful scenarios.

This section is primarily concerned with consideration of the important facts about a fire that ideally belong in a scenario. It is recognized that virtually all real fire investigations are handicapped by the absence of trained observers, especially at the early stages of the fire, so frequently one must guess what happened from fragmentary evidence. Nevertheless, it is useful to indicate what information is desirable. In some cases, one may want to: set up a simulation of a fire scenario to determine whether or not what one thinks happened could, in fact, really happen; instrument the fire and obtain quantitative data on critical fire dynamic elements; or investigate changes in the scenario from design modification and/or the substitution of different materials. Certainly, complete knowledge of the relevant factors is essential.

In the discussion of fire scenarios that follows, the physical behavior of the fire is emphasized, but its interactions with humans are deemphasized. This policy is observed because the committee's study is directed at fire safety via modifying materials rather than people. Nevertheless, it is obvious that people may enter into the fire scenario by: (1) starting or otherwise causing the fire, (2) preventing the fire, (3) detecting the fire, (4) extinguishing the fire, (5) escaping from the fire, or (6) being killed or injured by the fire.

3.3.2 Prefire Situation

In general, important events in the fire scenario occur long before the ignition source starts the fire. Frequently decisions made during planning, design, and manufacturing will profoundly affect the subsequent events in the fire chain; thereafter, it is essential that appropriate attention be directed toward the prefire situation since, in some instances, the optimum solution will result from action taken long before the fire begins.

Thus, the first step in the development of a fire scenario should include the gathering of data such as governing regulations, plans and specifications, manufacturer's records, inspection records, and operating records. Attention should be directed towards the rationale for material selection, how and where the materials were to be used, and how materials were installed. Specifically, one must know whether the materials meet the applicable specifications, whether they were used properly, and whether they were installed correctly. These and similar pieces of information are essential to the completeness of any fire scenario.

3.3.3 Ignition Source

In general, the fire starts with an ignition source. In perhaps a majority of cases, this initially is "wanted" combustion that leads to an "unwanted" fire. Examples are the discarded cigarette, the soldering torch, and the defective heater. Another large class of sources involves failures of electric circuitry and equipment or hydraulic systems. Numerous other special cases such as wayside power sources, tire blowouts or spontaneous combustion may be listed.

One often needs detailed information about the ignition source to characterize it quantitatively because in many cases the likelihood of ignition of the target fuel is marginal (e.g., a cigarette falls on the vehicle aisle carpeting but ultimately may self-extinguish, or a blowtorch impinges on the plastic coated panel for a second but only chars it). Unless the detailed characteristics of an ignition source are known, one cannot predict whether a fire will result.

The primary parameters of the ignition source are:

1. Maximum temperature.
2. Energy release rate.
3. Time of application to target.
4. Area on contact.

On a more sophisticated level, one may need to know the modes of heat transfer from the source to the target; this may be some combination of conduction, convection, and radiation. The degree of air motion or turbulence may influence spontaneous ignition of a heated vapor rising from a surface. Access to oxygen also is important (e.g., a target immersed in hot combustion products may not ignite because oxygen is excluded by the heat source itself).

The most important single fact to recognize about a potential ignition source is that, for solid polymers which are not readily ignitable, a "strong" igntion source generally will ignite the target while a "weak" one will not. The "strength" of the source depends on the energy flux, on the time of application to the target and, sometimes, simply on the product of these two.

3.3.4 Ignited Material

The first material to be ignited by the ignition source is generally crucial to the scenario. The question is: Given an exposure to an ignition source, how does the probability of ignition occurring depend on the properties of the target material? In most cases, the initial target material characteristics are vital in determining whether ignition occurs. Thus, a detailed description of the target material properties is vital to the scenario.

If the target material is a flammable liquid, its ignitability will depend on whether it is in the form of a stationary pool, a foam, a mist, or a spray. Assuming it is a stationary pool, its initial temperature is governing. If the temperature is below the fire point, ignition will occur only after sufficient heating has brought a substantial portion of the liquid to the fire point. If the initial temperature is above the fire point, ignition of the fuel vapors above the pool will occur immediately and the pool will easily sustain burning.

In most fire scenarios, the target material is solid. The ignitability of a solid depends not only on its chemical composition, but also on the energy balance at the surface (including radiation), its thickness and thermal properties, and on its configuration.

Under the heading of chemical composition of a target material, the following factors are especially relevant:

1. The basic material may contain small percentages of additives (fire retardants) or impurities that may have major effects on ignitability.
2. If the material is hygroscopic (e.g., cotton), the initial moisture content will vary over a wide range depending on prefire humidity and will have an important influence on ignitability.

3. If the material contains several major constituents (e.g., flexible polyvinyl chloride that contains a large proportion of plasticizer), the ignitability depends on the more volatile constituent.
4. The target frequently will be composite in nature, consisting of an outer skin material and an underlying material, either of which may contribute to ignitability.

The importance of the energy balance at the surface is shown by attempting to ignite a single piece of wood, e.g., a two-by-four. No self-sustained burning will result unless the ignition source is applied long enough to permit the average temperature of the wood to reach about 320°C. However, a match placed between two vertical two-by-fours close together will give self-sustaining burning. The single piece of thick wood cannot continue to burn because of the high rate of radiant energy loss from the charred hot surface to the cold surroundings. This effect is less important for materials that burn at lower surface temperatures, such as noncharring thermoplastics. Radiant input from the ignition source also can be important so the reflectivity of the target material is also a significant factor in such a case. This factor is of major importance in tunnel fires, in vehicle fires occurring near buildings, and, particularly, in the "canyon" type downtown area of cities.

The enhancing effect of radiant heat from a fire in a subway has been largely ignored. When considering the heat load of various plastics used in vehicle construction, one tends to think of the combustion energy dissipating into the environment; however, where there is a reflective wall, a significant amount of this energy can be returned adding to the heat flux, intensifying the heat, causing ignition of unburned material and raising the combustion temperature of the burning plastics. In two tunnel fires, one in the Montreal Subway in January 1974, and the other in the Boston Green Line on January 2, 1975, the concrete tunnel walls were said to have glowed red hot, and apparently contributed to the ignition of attached cars in the trains.

The thickness and thermal properties of a material are vital in determining the time required to achieve ignition when a given heat flux is applied to the surface. This relationship obviously becomes crucial in a scenario if the heat flux is of relatively short duration. A distinction must be made between "thermally thick" and "thermally thin" materials. The time to ignition for a "thermally thick" material is independent of the thickness and is controlled by the "thermal inertia," which is the product of the thermal conductivity and the heat capacity per unit volume. For a "thermally thin" material, the time to ignition is proportional to the product of thickness and heat capacity per unit volume (fabrics are generally in this category). Whether the material behaves in a "thermally thick" or a "thermally thin" manner depends not only on the thickness, but also on the heating rate, the heating time, and the "thermal diffusivity," which is the ratio of thermal conductivity to heat capacity per unit volume.

In the case of a thin flammable material (carpet, paneling, etc.) in thermal contact with an underlying material, the thermal properties of the underlying material can influence ignitability by the degree to which the underlying material acts as a heat sink.

The configuration of the target material also can be of great importance. The foregoing discussion has implied a one-dimensional geometry. In reality, ignition tends to occur more readily in a crevice or fold at an edge or corner than in the middle of a flat surface, and most vehicles have many such configurations.

3.3.5 Flaming or Smoldering Combustion

Some combustible materials may burn in either a smoldering mode, like a cigarette, or a flaming mode. Also, a material may smolder for a certain length of time and then spontaneously burst into flame.

In general, only solids with very low thermal conductivity, such as porous solids or thin solids not in contact with a heat sink (e.g., a suspended cotton thread or a free-standing piece of paper) can smolder; a seat cushion made of polyurethane and foam rubber under a synthetic fabric cover can burn in the smoldering mode. Smoldering is characterized by much lower fire spread rates than flaming combustion.

Smoldering is important in that: (1) the smoke or gases produced may permit detection of the fire at an early stage, (2) the pyrolysis products may be toxic, (3) smoldering removes moisture, and (4) a transition to flaming after a long period of smoldering may produce a very rapidly growing flaming fire because of the preheating of fuel and accumulation of combustible gases during the smoldering period (especially if a new source of oxygen is provided, as by opening a door).

It is known from the differing response characteristics of smoke detectors that smoke produced in flaming combustion is different from that produced by smoldering combustion of the same fuel. This difference must be taken into account when selecting smoke detectors. Smoldering may continue for a very long time (e.g., a seat cushion might smolder for hours) and scenario analysis therefore should consider the possibility of a long time lag between ignition and active flaming.

The burning of charcoal generally is referred to as glowing combustion rather than smoldering. The importance to the fire scenario is that cellulosic materials, after flaming combustion is finished, continue to glow for a substantial time as the residual charcoal is consumed. During this time, the possibility of a resurgence of the fire exists.

Also, when a gaseous extinguishing agent such as carbon dioxide or a halocarbon vapor is applied to a fire, it may stop the flaming combustion, but a smoldering combustion may continue (deepseated fire) and after a time the extinguishing vapor may dissipate and the flame rekindles. Thus, an automatic "one-shot" gaseous extinguishing system may not ensure protection unless the fire is held in check long enough for effective manual response.

3.3.6 Fire Spread

3.3.6.1 General

Unless a person is wearing or sleeping on the orignally ignited item, the fire is not apt to do much damage until it has grown by spreading some distance from the point of ignition. The rate of spread is very important in the scenario because it defines the time after ignition when the fire reaches a dangerous size. The "dangerous size" may relate either to the rate of generation of toxic and smoky products or to the difficulty of extinguishment. The ability to detect, fight, or escape from the fire depends on the time to reach a dangerous size and, hence, on the spread rate.

Fire may spread either from one continuous fuel element to the next or by jumping across a gap from the initially ignited material to a nearby combustible item. These two cases are discussed separately.

3.3.6.2 Fire Spread Over the Initially Ignited Material

The rate of flame spread over a solid surface in the horizontal or downward direction is often quite slow, sometimes as little as 1 inch per minute. However, if the material is "thermally thin" or has been preheated by radiation or convection from hot combustion products, the flame can spread quite rapidly. If the fuel is so arranged that upward propagation can occur, it will occur very rapidly and at a progressively accelerating rate. If the fuel is arrayed as a lining of a vehicle wall or ventilation duct, with the air supply coming from the left, for example, and the combustion products exiting on the right, the fire will spread rapidly from left to right until it penetrates sufficiently far into the duct so that the oxygen is exhausted. It then will stop spreading until the originally burning fuel is consumed, after which the fire will move down-stream. Thus, in this situation, the effects of ventilation is controlling.

Indeed, for any fire burning in a compartment with limited air supply, the rate of spread will decrease as the air becomes vitiated by combustion products. However, spontaneous breaking of windows or deliberate actions of firefighters to improve visibility by ventilating the fire will have an accelerating effect on spread rate. Most vehicles will have a "ventilated" type fire.

Fire spread over a liquid is relatively slow when the liquid is well below its flash point but possibly a hundred times as rapid if the liquid is above its flash point. For liquids below their flash point, motion within the liquid induced by the fire is important in determining the spread rate.

3.3.6.3 Fire Spread to Secondary Material

If an originally burning material is separated by a gap from the nearest secondary combustible and the flame does not impinge directly on this secondary material, the fire will die out after the original material is consumed unless it can spread across the gap by some mode. For example, the fire may radiate directly on the target or may convectively heat the vehicle ceiling and upper walls, which then radiate on to the target. If the vehicle is burning in a tunnel, the tunnel walls will radiate heat; in fact, the tunnel walls may glow red hot. Hot smoky gases accumulating under a ceiling may radiate on to the target. In some instances, a combination of these effects may occur.

In any case, the effect of the radiation is to preheat the secondary material until it pyrolyzes, emitting flammable vapors. At this point two possibilities exist. Either the secondary surface may ignite or a sufficient concentration of a flammable vapor mixture is achieved so that the original flame may spread through this vapor cloud to the secondary material.

Other modes exist to force fire spread across a gap. If the original burning object is a thermoplastic, it will melt, and burning droplets may fall and ignite secondary fuels they may encounter.

3.3.7 Evolution of Smoke and Toxic Gases

3.3.7.1 General

Smoke and toxic gases are important to the fire scenario in at least three ways. First, they may provide a means of early detection of the fire. Second, smoke interferes with visibility and, hence, with escape or with firefighting. Third, smoke and toxic gases have psychological and physiological effects on humans, including confused thinking, incapacitation, and death. In many instances, death in vehicle fires is a result of the toxic combustion products and not a result of the heat and flames from the fire.

In addition, other aspects of smoke and fire gases may exist. The smoke may interfere with some critical operations such as driving a bus. The smoke may be important in the fire spread process by virtue of its radiation emission or absorption. Moreover, substantial property damage may be caused by smoke or corrosive combustion products.

3.3.7.2 Automatic Detection

The first consideration in automatic detection is the rate of smoke movement from the fire source to the detector. Under a no-fire condition, the air movement in a vehicle is determined by any existing forced convection for heating, air conditioning, or odor-removal purposes; or by open windows along with external wind conditions; or by free-convective motions driven by heat sources. For very small fires, the buoyancy effect of the fire heat will be negligible and the smoke will follow the existing air circulation paths. When the fire becomes larger than some critical size, the hot fire plume will rise to the ceiling and then flow under the ceiling, creating an entirely new circulation path in the passenger/cargo space. If, before the fire, the upper portion of the space is warmer than the lower portion, as is often the case when the vehicle is stopped, temperature-induced stratification may exist and smoke may rise halfway up and then spread laterally. For early detection of fires, particularly when vehicles are stored in a garage or lot, the foregoing factors are crucial in determining detector response.

The next consideration is the response characteristics of the automatic detector to the smoke. Such characteristics vary with the time-dependent concentration and particle size of the smoke at the detector, the velocity of the smoke passing the detector, orientation of the detector to the flow, smoke entry characteristics of the detector chamber, and the operating principle of the detector (optical transmission, optical scattering, ionization), as well as the sensitivity setting of the detector circuit battery voltage, etc. It is especially important to note that different combustibles, or the same combustibles flaming or smoldering under different ventilation conditions, produce smoke of different particle size and, hence, detection characteristics. It is also known that the smoke may "age" after it is formed (i.e., the agglomeration of smaller particles into larger ones will occur) with consequent effect on ease of detection.

3.3.7.3 Visibility

The optical scattering and properties of the smoke depend strongly on particle size as well as concentrations so the vision-obscuring aspects which interfere with escape or firefighting are strongly dependent on the type of combustible and mode of combustion. For example, incomplete burning of polystyrene or rubber produces large soot particles capable of obscuring vision even at low concentrations. The lachrymatory effects of gases such as aldehydes or acids associated with the smoke particles also have been shown to be important in interfering with vision.

3.3.7.4 Toxic Effects

Smoke and toxic gases are sometimes more important than heat and flame as a cause of death in vehicle fires. Carbon monoxide is the chief toxicant according to present knowledge; however, other specific substances that may be present in the smoke (e.g., acrolein, HCN, HCl, HF, and CO_2) may be very important in certain cases and may have synergistic effects.

The critical survivable concentration of toxicant depends on the time of exposure, which, when escape is extremely difficult (as in a subway tunnel under water), depends on the history of the fire as discussed in other sections. Also, combined effects of toxicants with heat, excitement, loss of vision, etc., are believed important in determining survival as is the original condition of health of the subject and previous intake of alcohol or drugs. Confused mental processes induced by toxicants may be of critical importance to survival in cases where the subject has to make a rapid or critical decision on proper escape tactics.

3.3.8 Extinguishment

At some point in the development of each scenario, either manual or automatic extinguishment activity may commence. This may involve smothering the fire or applying water or some other agent (e.g., Halon 1301). The techniques of extinguishment are outside the scope of this study; however, the effectiveness of this extinguishment will depend on the burning characteristics of the polymeric combustible. If the fire has become too large or is growing too rapidly at the time extinguishment is attempted, the fire will not be controlled.

Accordingly, the rate of fire spread and the maximum rate of burning of the fully involved combustible are important parameters. For manual firefighting, the critical questions are: How closely can the fire fighter approach the fire and will smoke prevent him from determining where the fire is? If he has a hand-held extinguisher of given capacity, will it be enough to do the job? When automatic dispensers are present, there is generally no problem unless the fire is shielded from the dispensers (i.e., in a closed compartment) or unless it is a high-intensity fire.

3.3.9 Flashover

Flashover is a critical transition phase of a fire in a vehicle. In general, it will be a ventilated compartment since otherwise the fire will tend to smother itself before the flashover stage is reached. Prior to flashover, a local fire is burning in the compartment, the rate of which is determined by the extent of fire spread to that time. After flashover, all flammable contents of the compartment are burning or rapidly pyrolyzing, flames are projecting out the doors or windows, and the burning rate within the compartment is determined by the rate of ventilation and/or the total exposed fuel area. Flashover often occurs suddenly, within seconds, and is characterized by very rapid fire spread throughout the compartment with flames violently rushing out the doors or windows.

Whether flashover can occur in a compartment depends on its size and shape, the ventilation available, the intensity of the initial fire, and the quantity and disposition of secondary fuels. If flashover can occur, the time required for its occurrence will depend on the foregoing variables plus a thermal inertia of the space with the ceiling height being particularly important.

In the pre-flashover period, the upper portion of the compartment is filled with hot, smoky, oxygen-deficient gases. The lower portion contains relatively cool, clean air coming from the doors, windows, or passenger ventilation ducts. At some intermediate level, perhaps two feet under the ceiling, there may be both sufficient oxygen and sufficient heat so that target fuels at this height could readily ignite. Decorative side panels or advertising panels are examples of polymeric materials in this region.

Radiation is probably of major importance in flashover. Thus, infrared emission, absorption, and reflection characteristics of objects and smoke in the compartments are highly relevant.

The larger the volume of a compartment, the less likely it is that a fire of given size will cause flashover. Data on simulated fires indicates that, for a 12 foot by 12 foot by 8 foot compartment, a fire consuming 2 pounds of fuel per minute could produce flashover in about 20 minutes while if the combustion rate were twice as high (i.e., 4 pounds per minute), flashover would occur in 1.5 minutes. One would suspect that even a very large, sparsely furnished compartment would flashover if a high rate of initial burning is achieved, because the time of flashover is extremely sensitive to rate of heat release. The Montreal subway fire of 1974 illustrates a fire in an enclosed space (tunnel) with low fuel loading but presumably very high heat release rate because of the plastic foam padding on seats.

3.3.10 Spread to Adjacent Compartments and Catastrophic Failure

Compartmented vehicles are sometimes designed with the expectation that a fire in any one compartment will be confined by the structure itself so that either the fire is extinguished or the fuel is exhausted before the fire breaks through. Thus, the scenario should include information on the fire endurance rating of the relevant structural elements.

Fuel loading influences the duration of a fire once it has grown large enough to become ventilation-controlled. The fire load may range from a few pounds per square foot in austere vehicles to an order-of-magnitude higher fire load in plush occupancies such as the new WMTA Metro subway and Metro bus systems. If the fuel is primarily a polyolefin or rubber, the stoichiometric air requirement will be up to three times as large as if the fuel were primarily cellulosic and the heat released per pound of fuel also will be much higher, and a ventilation-limited polyolefin fire may burn differently than cellulosic fire.

If the fire compartment has openings to other sections of the vehicle (e.g., open doorways, ventilating ducts, improperly fire stopped or inadequately sealed openings in floors and walls), these become critical elements in the scenario. Even if the fire itself is confined to the compartment of the origin, the spread of smoke and toxic gases throughout the vehicle and its associated system would have catastrophic effects (e.g., a burning bus in a traffic-filled tunnel). Therefore, the presence of materials that yield significant quantities of smoke and toxic gases is a critical element of the scenario.

If the structural elements of the vehicle can fail as a result of heating during a fire (as is often the case with aluminum skin composites), collapse of the structure may occur. Thus, the thickness and integrity of insulation on structural elements as well as the elements making up the strength members become important to the fire scenario.

3.3.11 Spread to Other Vehicles

If a vehicle becomes completely involved with fire, a substantial probability exists that adjacent vehicles will ignite (as when buses are stored close together). Ultimately, a conflagration involving a large area may result. Propagation could occur either by radiation or convection and would be supported downwind.

Potentially critical factors in the fire scenario are: magnitude and direction of the wind, separation distance between vehicles; ignitability by radiation of material inside windows facing the fire, combustible trash between vehicles, and propulsion of burning debris from explosive flashover or fuel explosions.

Complete involvement of a vehicle usually occurs in a sufficiently late stage of fire so that firefighters generally will be present. Their tactics in wetting down adjacent vehicles are extremely valuable in preventing spread to other vehicles. Conversely, if the fire is simultaneously burning in many areas, as could be the case for a fire caused by civil disorders or military incendiary attack, firefighting probably will be inadequate, and the degree of spread will depend on the intrinsic "hardness" of the vehicles involved.

3.3.12 Essential Fire Scenario Elements

A scenario should cover as many as possible of the following points:

1. The prefire situation.
2. The source of ignition energy (described in quantitative terms).
3. The first material ignited (characterized as to chemical and physical properties).
4. Other materials that play a significant role in the growth of the fire.
5. The path and mechanism of fire growth (particular attention should be given to fuel element location and orientation, ventilation, compartmentation, and other factors that affect fire spread).
6. The possible roles of smoke and toxic gases in detection, fire spread, and casualty production.

7. The possibility of smoldering combustion as a factor in the fire.
8. The means of detection, the time of detection, and the state of the fire at the time of detection.
9. Defensive actions taken and their effect on the fire, on the occupants, and on other factors.
10. Interactions between the occupants of the vehicle and other system elements and the fire.
11. The time sequence of events from the first occurrence of the ignition energy flux to the final resolution of the fire.
12. Vehicle or system design that enhances or retards the spread of the fire.

The scenario should permit generalization from the particular incident described. It should provide a basis for exploration of alternative paths of fire initiation and growth and for analysis of the effect on fire safety performance of changes in materials, design, and operating procedures. When used in this way, the fire scenario can be an effective tool in increasing fire safety by increasing man's capability to visualize and comprehend the events.

3.3.13 Analysis of the Fire Scenarios

3.3.13.1 General

Prevention and control are the prime purposes of any fire scenario analysis. The merits of a comprehensive, factual analysis rests on the accuracy, level of detail, and completeness of the scenario; however, developing a fully useful fire scenario requires either a completely documented report of a detailed post-accident investigation and analysis specifically designed to determine how and where the fire started and progressed until extinguishment, or a similar report of a fully instrumented full-scale test burn, or a combination of both. In any case, until existing knowledge of the dynamics of actual fires is augmented by additional fire dynamics research, development of fire scenarios will be an art rather a scientific discipline. Nevertheless, the application of fire scenario analysis appears to be a most productive methodology to identify economical, effective means to improve vehicle fire safety in an increasingly complex environment.

The analysis of a fire scenario might be accomplished in various ways. One effective means would be to ask a series of questions concerning each essential fire scenario element. The answers to these questions should suggest means for prevention and control while providing a basis for materials selection, design criteria, validation of test methods, and promulgation of codes and standards as well as research and development objectives. Typical questions that one might ask are presented below.

3.3.13.2 Prefire Situation

1. Were existing regulations, specifications, and codes adhered to? Were they adequate? If not, why? Would they have been effective had they been enforced?
2. Was the vehicle designed to minimize fire spread (or to control/contain it)?
3. Were materials installed properly? If so, did they contribute to fire growth or did they help contain the fire?

3.3.13.3 Ignition Source

1. In as much detail as possible, what was the ignition source?
2. For how long was it in contact with the ignited material prior to flaming ignition? If this time is not known, could it be determined by a separate experiment?
3. Could the ignition source be eliminated? How (by education, by design)?

3.3.13.4 Ignited Material

1. What was the originally ignited material? If a composite, what were the various layers?
2. What was the application of the material (e.g., seat cover, carpet, cushion)?
3. How was it located relative to ceiling and nearest wall?
4. What were the ventilation conditions in the compartment?
5. Did melting and dripping of the ignited material occur? Did these occurrences significantly affect the fire spread?
6. Did the ignited material collapse, drop, etc.? If so, what effect did this occurrence have on the fire scenario?
7. Once the ignited object was fully involved, is it possible to estimate its heat release rate (in energy units)?
8. Are there other materials that could have been employed in this application which would not have ignited under the same conditions? If so, why were they not employed?
9. Did flammability tests on materials intended for this application adequately measure ignition resistance to this level of ignition source? Should they?

3.3.13.5 Flaming or Smoldering Combustion

1. Is it known that smoldering preceded flaming? For how long?
2. If unknown, was the ignited material capable of smoldering?
3. Can the volume of gases produced by smoldering be estimated?
4. Can the time-dependent concentration of smoke and toxic gases arriving at a strategic location some distance from the fire be estimated?

3.3.13.6 Fire Spread

1. How long did it take for the first ignited object to become fully involved?
2. If flame spread to a second object, what was the mechanism of energy transfer?
3. How was flame spread influenced by events such as the breaking of windows, opening of doors, and spattering of burning droplets?
4. Did one or two materials significantly control the fire spread route? Could the substitution of different materials or the incorporation of design modification alter the rate of fire spread and growth?

3.3.13.7 Smoke and Toxic Gases

1. If a smoke detector was present, how was it located relative to the fire? Did it res-

pond as expected?

2. If no smoke detector was present, how much sooner would the fire have been detected if it had been present in a logical location? Would such detection have been soon enough to make a crucial difference?
3. Was visibility obscured in an escape route? When did this obscuration occur relative to detection time? Which materials seem to contribute significantly to visibility obscuration?
4. Were victims affected by toxic substances?
5. What toxic substances caused death? Was there an autopsy?
6. Did toxic substances interfere with escape by contributing to confused thinking or decision making?
7. Could these toxic substances be attributed to any one material?
8. Did victims have pre-existing conditions such as limited mobility, circulatory disease, recent alcohol or drug intake, etc.?

3.3.13.8 Extinguishment

1. How large was the fire when first detected? What were visibility conditions at this time?
2. How much time elapsed between detection and attempted extinguishment?
3. How large was the fire when extinguishment was attempted?
4. What was the extinguishment technique and how successful was it?
5. If automatic extinguishers had been present, how much sooner would they have been expected to control the fire, and how much less might the loss have been?

3.3.13.9 Flashover

1. Did flashover occur? How long after ignition? How long after detection?
2. Can crucial elements in the fire growth and spread process be identified in relation to flashover?

3.3.13.10 Postflashover

1. Did the fire spread beyond the initial compartment? How? Was a door open?
2. How was the ventilation system involved in fire spread?
3. Did floor, walls, etc., fail? If so, after how long?
4. Did vehicle structure collapse occur? Was this collapse due to faulty specification, design or some other cause?

3.3.13.11 Spread to Other Vehicles

1 Did other vehicles ignite? Were they part of the same train? How far away were they? What material ignited first?
2. Was radiation responsible? Was there direct flame contact?
3. What were wind conditions?
4. Did firefighters attempt to protect exposed vehicles? How soon before spread occurred?

3.3.14 Summary

The availability of accurate, detailed, complete fire scenarios allows opportunity for an in-depth fire hazard analysis as illustrated by the spectrum of questions shown above. As such questions are raised and some answered, means for fire prevention and control will emerge. These means may involve better materials selection, more education, improved designs, installation of detection equipment, and more stringent specifications; however, whatever solution emerges, it will be based on a comprehensive overall system analysis of the problem.

3.4 A Spectrum of Fire Scenarios Involving Vehicles

3.4.1 Introduction

No two fires are alike in all their details. But, as was indicated above, all fires have certain elements in common that permit their systematic study and lead to generalized rules for increased fire safety. All fires have a cause, their initial growth is determined by physical parameters of the system, and these parameters together with external control measures intervene to limit the growth of the fire and determine the extent of loss. In this section, a variety of brief scenarios based on real fire incidents in which polymeric materials played a significant role is presented to illustrate the diversity of fires and to show the commonality that permits a scientific approach to fire safety. In preparing these illustrative scenarios, attempts were made to address: prefire environment, ignition source, material first ignited, other significant materials involved, fire dynamics, method of detection, extinguishment, and extent of loss.

3.4.2 Vehicle Fires

3.4.2.1 Scenario 1

A father had picked up his son at college at the end of the spring term. The son's belongings were loaded in the back of the station wagon. As they were driving down an interstate highway, the odor of smoke was detected. At first it was faint and was assumed to be from the heavy traffic. It grew stronger and a faint haze became visible in the rear view mirror. At this point they pulled up on the shoulder and opened the back of the station wagon to investigate. Smoke was seen rising from a foam rubber pillow that had been placed on top of other items. It was removed from the car and torn open, whereupon it burst into flames and was totally consumed at the roadside. The interior and contents of the car were saturated with the odor of burning rubber and required extensive cleaning. There was no other damage. The sun, shining in the curved rear window of the station wagon during a long stretch of straight road, had heated the foam rubber to the point at which auto-oxidation started and heat built up within the well-insulated interior. When the pillow was removed from the car, the smoldering foam rubber was exposed to a fresh supply of air and open flaming resulted.

3.4.2.2 Scenario 2

Children were left in a locked automobile while the parents went shopping. The children, playing in the back seat, found a pack of matches. The upholstery materials in the back seat became ignited, and then the foam plastic padding material became involved. Rescue was attempted by several passersby; however, the heat was too intense for rescue. The children died of burns and smoke inhalation before the fire department arrived.

3.4.2.3 Scenario 3

The bus driver had just parked his new bus at a turn-around spot after unloading his passengers approximately two blocks before. He smelled smoke and turned in his seat, and saw smoke coming from the rear of the vehicle. He attempted to extinguish the fire with a portable extinguisher but was driven back by the intense smoke. The foam plastic seating was the initial material observed to be on fire. Shortly thereafter the vinyl wall coverings and synthetic plastic windows became involved. Flashover occured in less than 3 minutes after discovery of the fire and the entire bus interior became involved. The entire interior was destroyed in less than 5 minutes. This was before the fire department could extinguish the fire. Fire officials investigated the fire in the bus and concluded that arson was the probable cause of the fire.

3.4.2.4 Scenario 4

As a subway train approached an underground station, arcing occurred between a blown-out steel-reinforced rubber radial tire in intermittent contact with the live third rail. This intermittent arcing continued; the train tires and hydraulic lines in the undercarriage caught fire. After these components ignited, the fire penetrated the plywood car floor and proceeded to burn the cabin interior, which consisted of plastic interior finish and flexible polyurethane foam seating. Passengers were removed prior to the burnthrough and a train man attempted to extinguish the fire by directing a portable carbon dioxide extinguisher at the flames. His efforts were unsuccessful and the fire was soon out of control. Firefighters were unable to approach the fire for over 2 hours due to the smoke and intense heat. Four cars of ten in the train were completely destroyed.

3.4.2.5 Scenario 5

A disk brake on the underside of a modern subway car overheated and caused heating of the aluminum plate that served as a barrier to external penetration into the passenger area. The car was unoccupied at the time. The aluminum plate covered a foamed plastic insulation material that became heated and then ignited. Open flaming occurred and was intensified by hydraulic fluid from lines which served the brake. The fire was witnessed by passersby when the train was above the ground. The fire department was called and met the train at the next convenient area. The intervening period, the fire had broken through the undercarriage and had caused the ignition of the vehicle's seat cushions (foamed plastic) and fibrous glass reinforced plastic lining materials. Firemen wearing breathing apparatus entered the car and quickly extinguished the fire with hand lines. According to fire officials, the car appeared to have been on the verge of flashover when the fire department arrived.

3.4.2.6 Scenario 6

An electrical malfunction in a trolley car in a subway caused the trolley wire to separate and dangle, unnoticed and live. The following train of three trolley cars, carrying almost 400 passengers, struck the 600-volt line which arced and jumped onto the roof where it made an 18-inch hole through the roof metal and ignited sound-deadening material in the ceiling. Because of confusion, there was a delay of 20 minutes in calling the city fire department. The passengers had been evacuated safely, but the firefighters encountered intense smoke. Lack of ventilation systems and standpipes in the tunnel made firefighting particularly difficult. The fire eventually was extinguished but not before the first car was completely gutted and the second car damaged by radiant heating from the walls of the tunnel. Thirty-four firefighters were hospitalized for inhalation of smoke from burning plastics.

3.5 Analysis of Specific Fire Scenarios

3.5.1 Introduction

This section is primarily intended to illustrate the fire scenaro approach to improved vehicle fire safety. To this end, two generalized fire scenarios are developed — one will deal with a bus fire and the other with a commuter car fire. These scenaros include the essential fire elements (e.g., ignition source and first material ignited) identified in Section 3.3 and the scenarios are prepared to permit generalization from the particular incident described.

The major reason for the development of fire scenarios is to identify causes of action that will help to prevent and control similar fires in the future. Thus, these scenarios are analyzed to identify specific hazards, suggest opportunities for fire prevention, and direct attention towards methods for control. The analyst also should question the adequacy of specific materials, design approaches, and regulatory codes and procedures.

The specific goal of this section is to demonstrate that fire scenario development and analysis is a productive methodology for improving the selection and use of polymeric materials to increase fire safety in a transportation system.

3.5.2 Scenario — A Bus Fire on the Highway

3.5.2.1 Description

A metropolitan mass transit bus was proceeding in an express bus lane at highway speeds during the evening rush hour. A passenger called the driver's attention to flames coming up from the left rear wheel. The fire burned through the plastic wheel well cover and into the bus interior. The bus was stopped and all 70 passengers safely evacuated. Heavy smoke and gases developed within the bus as the seats, paneling, and other polymers became involved. Fire extinguishers applied by the drivers of two other nearby buses failed to control or even slow the fire. The bus burned totally in the middle of the highway. (Figure 1).

3.5.2.2 Analysis

3.5.2.2.1 Prefire Considerations

Data relative to the bus was gathered, including plans and specifications, manufacturing records, and operation and maintenance records. Examination revealed that the bus was well built and had been well maintained and properly operated.

3.5.2.2.2 Ignition Source

Statements by bus passengers and occupants of nearby automobiles identified the ignition source as being in the left rear wheel assembly. Examination of the assembly revealed that a wheel bearing had burned out providing sufficient heat to ignite the axle grease. Further analysis suggested that overheated brakes or under-inflated tires could have been equally serious ignition sources; mass transit buses operating at highway speeds have several potentially serious ignition sources in their wheel assemblies.

3.5.2.2.3 Ignited Material

The target material was the axle grease; from there the fire progressed to the tires and then through the composite polymer wheel well cover into the passenger compartment. The fire then ignited the polyurethane foam seat cushions and other polymers in the bus. The characteristics of these materials were crucial to the development of the

Figure 1. Bus Fire.

fire; analysis revealed that the fire properties of these materials were such as to support growth of the fire. Clearly, a "fire-stop" material was needed between the wheel assembly and the passenger compartment; other design changes to reduce the fire vulnerability of the wheel assemblage were also indicated.

3.5.2.2.4 Flaming or Smoldering Combustion

The strong flames, burning through the wheel well cover, initiated flaming combustion of the bus interior without smoldering. The fire proceeded to destroy the bus. Analysis also revealed that smoldering combustion, perhaps started by a cigarette in a seat crease, although taking a longer period to develop, could equally well develop into full flaming combustion and loss of the bus. The materials used and the design of the bus were such that active quenching (by human or automatic means) would be required to stop the fire sequence in either the flaming or smoldering mode of combustion. Serious doubts must be raised about design and material selection under such circumstances; better materials are available.

3.5.2.2.5 Fire Spread

The fire spread rapidly from the ignition source, involving all polymers in its path. With poor materials (from a fire safety standpoint) and no passive fire containment in the design, loss of the bus was probable. This situation is prevalent in mass transit buses in the United States; almost all exhibit these undesirable characteristics. Better materials (from a fire safety standpoint) should be used.

3.5.2.2.6 Evolution of Smoke and Toxic Gases

Large amounts of smoke and toxic gases evolved during the flaming combustion. The same materials would have produced large amounts of smoke and toxic gases (perhaps of different species and character) under smoldering combustion. While this matter was not serious in this particular case (on a highway) and all the passengers did escape, the problems of smoke and toxicity in a tunnel or on a crowded downtown street could add a significant additional hazard to public safety. The smoke and toxic gas generation characteristics of the bus materials thus become a major matter for analysis in this *and all other* bus fire scenarios.

3.5.2.2.7 Automatic Detection

The bus was fitted with automatic detection devices in the engine compartment, but there were none in the passenger compartment. In this scenario, the lack of automatic detection was not an important factor. In a smoldering fire (e.g., when fire starts from a cigarette smoldering on a seat in an unoccupied section of the bus), detection may be vital to survival.

3.5.2.2.8 Extinguishment

The manually applied CO_2 extinguishers were inadequate for this intense fire. Because of poor material choice and design, once the fire reached the passenger compartment, fire control would be very difficult, thus raising the question of passive versus active fire containment philosophies. Analysis of scenarios of passenger compartment fires of similar buses leads to the conclusion that, no matter what the ignition source and ignited material, there is a high probability of the bus interior being gutted despite active fire control measures. This conclusion raises not only the issue of better material selection, but also better design for fire safety, including reduction of the fire load, designed fire stops, fire control procedures and operations, early detection, etc.

3.5.2.2.9 Flashover

Flashover (see Section 3.3.9) is usually a critical point in bus fires. After flashover occurs, there is little likelihood of saving the bus; more importantly, any passengers still on the bus at the time of flashover will be seriously burned or become fatalities. Flashover can occur within a few minutes of fire initiation. Although flashover did not occur in this scenario, it is a possibility in most fires; analysis must include determination of flashover occurrence (or the probability thereof).

3.5.2.2.10 Spread to Other Structures and Vehicles

Fortunately, the open highway permitted other vehicles to give the burning bus a wide berth; there were no nearby structures. In a tunnel or narrow downtown street, spread of the fire can be a serious safety hazard (see Section 8.1.4.2.9).

3.5.2.2.11 Fire Load

The bus interior contains many polymers that will burn and emit smoke and toxic gases. The most serious contributors from a flammability and total heat load standpoint are the polyurethane foam seat cushions. Wall and ceiling panels, insulation, carpeting, and flooring also contribute to the fire load (although not as heavily). Scenario analysis indicates that, from a material standpoint, the urethane foam should be replaced by better materials (such as neoprene foam), that flooring (structure) should have excellent fire safety performance, that thermal and noise reducing insulation must have satisfactory fire resistance, and that a number of less vital material changes are required. From a design standpoint, there is a need for passive fire containment, fire stops, and system operations to minimize fire spread and fire effects (e.g., ventilation control during fires and elimination of smoke and toxic gas distribution through the passenger compartment).

Analysis of other bus fire scenarios with other ignition sources, other paths of fire spread, and other final results yields many of the same recommendations (see 8.1.4 ff).

3.5.2.2.12 Passenger Egress

All passengers were active and agile; all escaped. Handicapped persons might have had difficulty in a smoke and a gaseous atmosphere, thus, pointing to selection of materials with reduced ignition proclivities and lower flame spread rates to maximize escape time.

3.5.2.3 Summary

Section 3.5.2.2.1 through 3.5.2.2.12 analyze a specific fire scenario providing data pertinent to the fire and, more importantly, guidance as to what needs to be done to improve the fire safety of mass transit buses through better design and selection of better materials.

By application of this method to other bus fires, it would be possible to develop an optimum course of action to improve all buses and to develop the trade-offs between risks and costs to prioritize the application of available assets to the fire safety problem.

3.5.3 Scenario — A Self-Propelled Commuter Car Fire in Suburbia

3.5.3.1 Description

The first car of a three-car commuter train collided with a truck at a crossing. The 40-gallon gasoline tank of the truck exploded and subsequently ignited the truck load of corrugated cartons. The right front corner of the commuter car, including the motorman's step trap door was distorted and allowed intense flames to penetrate the commuter car via the outside door, burning through the neoprene gaskets and breaking the glass window.

In his hasty exit from the control booth, the motorman left the door open into the passenger compartment. The fire entered the forward section of the car (quickly and safely evacuated by 40 passengers).

The flames burned out the forward section. The fire department arrived 5 minutes after the crash and put out the fire in time to save the rear section.

This scenario was analyzed using the same procedure as for the bus fire in Section 3.5.2.

3.5.3.2 Analysis

3.5.3.2.1 Prefire Considerations

Data relative to the car were gathered, including plans and specifications, manufacturer's records and procedures, inspection records, and operating records. These data indicated that all materials met or exceeded specification and regulation requirements and had been properly maintained.

3.5.3.2.2 Ignition Source

The ignition source was the heat energy developed from the crash.

3.5.3.2.3 Ignition Material

In this case the truck fuel, exterior to the car, was initially ignited. The flames spread to the interior of the car because the motorman's door was left open; had it been closed, the time available for escape would have been increased and the severity of damage would have been reduced.

The analysis indicates the value of an automatic closure device on the motorman's door.

3.5.3.2.4 Flaming or Smoldering Combustion

The intense heat from the burning truck fuel ignited the forward section furnishings without smoldering taking place.

3.5.3.2.5 Fire Spread

The rapid spread of the fire in the forward section was due primarily to the large ignition source and the ignitable material in the forward compartment. It was significant that, due to car design, the fire did not readily spread into the car's rear compartment, which was almost undamaged by the fire.

3.5.3.2.6 Evolution of Smoke and Toxic Gases

Although heavy black smoke and some toxic gases were generated in the fire, they were not a significant factor in passenger egress or safety. However, the air conditioning system sucked smoke and gas into a plenum running the length of the car and distributed the smoke throughout the car. A "fail-closed" closure in the air conditioning system or a "fail-off" air conditioning system would have prevented this undesirable smoke and gas distribution.

3.5.3.2.7 Automatic Detection

No detection devices were fitted; absence of detection devices did not affect this scenario.

3.5.3.2.8 Extinguishment

The fire department, which arrived within 5 minutes of the crash, was able to prevent the spread of the fire to nearby grass and structures.

3.5.3.2.9 Flashover

Flashover was not involved.

3.5.3.2.10 Commuter Car Design

The vehicle was designed with two identical sections separated by a vestibule containing two pairs of seats on each side with partial walls separating the forward and rear sections. This design had the effect of partially restricting fire spread from the forward section.

The seat cushions were neoprene foam covered with PVC fabric. Wall and ceiling interior panels were melamine. Windows were polycarbonate. The undercarriage, floor, walls, and ceiling structure were steel (some stainless).

Considering the intensity of the fire and the large fuel load, it was remarkable that the rear section of the car did not catch fire and burn (as have cars of different construction under similar fire conditions).

The analysis indicated that the design of the car and the selection of optimum materials, particularly in the structure and for the seats and windows, prevented complete destruction of the car. (This car was refurbished and returned to service.)

3.5.3.3 Summary

The analysis of this fire scenario amply demonstrates the value of using highly fire-retardant materials and designing a vehicle to restrict the progress of the fire. The passengers were allowed sufficient time to evacuate the vehicle, and the fire progressed slowly enough to give the firefighters time to arrive and save a refurbishable vehicle.

In addition, although not significant in this fire, other fire safety factors emerged from the scenario analysis (i.e., automatic door closure for the motorman's door and "fail-safe" air conditioning system modifications).

3.5.4 Evaluation of Scenario Analysis

In Section 3.5.2 a fire scenario analysis was presented which demonstrated the consequences of poor material selection, design errors and installation faults. In Section 3.5.3 a scenario analysis was presented demonstrating the benefits of good material selection and good design. (Chapters 7, 8, and 9 also contain abbreviated scenarios and analyses.)

Scenario analysis is a powerful, underutilized tool that can improve fire safety. Like the "case method" in business studies, it permits broad or narrow focus of many experts on a well-defined (perhaps hypothetical) situation.

Scenario analysis consistently confirms that fire efforts are, unfortunately, largely directed to items or components when, in fact, the problems are those of complex systems. Polymeric materials are major elements in these complex systems; new materials are being developed and introduced without full understanding of the potential consequences of their use, particularly as they relate to fire safety performance. Scenario analysis can assist in better defining the roles that materials can play and what consequences might result. Scenario analysis appears to offer a superior tool in determining the best course for increased fire safety in an increasingly complex multisystems society.

Compilation and review of the analyses of many vehicle fire scenarios would provide an improved basis for developing greater fire safety. Specifically, the analyses would result in: selection of improved materials for use in vehicles; development of improved small-scale tests for vehicle materials, components, subsystems, and systems; improved vehicle design criteria and methods; provision of a basis for necessary, credible full-scale fire tests for various types of vehicles; development of improved vehicle operating and maintenance procedures; and development of risk assessment and trade-off methodology needed but not now available.

3.6 Conclusions and Recommendations

Conclusion: Despite serious deficiencies, current knowledge of fire dynamics is sufficiently advanced, if used in depth, to assist in improving vehicle fire safety, developing new and more useful fire tests, and predicting vehicle fire performance. To date, fire dynamics knowledge seldom has been used to assess the performance of land transport vehicles and associated systems.

Conclusion: The development and analysis of a fire scenario leads to the identification of critical stages in fire development, suggests opportunities for fire prevention, and directs attention towards various methods for fire control.

Recommendation: Increase use of current fire dynamics knowledge in mass transit vehicle design to prevent hazards and reduce losses.

Recommendation: Develop a wide spectrum of vehicle fire scenarios and quantify specific fire dynamics elements in these scenarios (e.g., fire spread and heat release rates) by information obtained from vehicle fire experiments.

Recommendation: Prepare vehicle scenarios to permit generalization from the particular incident described and to provide the basis for exploration of alternative paths of the fire initiation and growth and for the analysis of the effects on fire performance of changes in materials, design, and operating procedures.

Recommendation: Train vehicle design engineers and transportation systems fire safety personnel in the development and use of fire scenarios to enable them to readily identify critical fire hazard elements and appropriate safety measures.

CHAPTER 4

POLYMERIC MATERIALS

4.1 General Considerations

This chapter provides broad general information on the state of the art of the fire safety aspects of the main classes of natural synthetic polymers as they are used in transportation vehicles.

4.1.1 Fire Safety Aspects of Polymeric Materials

It must be emphasized that the optimum selection of material from a safety standpoint is a difficult task even for the "expert" because of the many competing design elements that must be considered before a proper selection can be made. Materials selection is a primary way to reduce the threat of fire. General fire-consciousness in system design, use of structural materials with improved fire safety characteristics, fire detection, and firefighting are equally important aspects of coping with fire hazards. They must be viewed together and analyzed in a systems approach to the problem.

Consideration of polymeric materials with improved fire safety characteristics is a process in which competing fire safety requirements must be reconciled. For example, use of a fire retardant additive that leads to a decrease in ease of ignition or flame spread but which also leads to an increase in the production of smoke or toxic combustion products might not be tolerable.

4.1.2 Approaches to Improving the Fire Safety Characteristics of Polymeric Materials

No organic polymeric material can withstand intense and prolonged heat without degradation even in the absence of oxygen. Given sufficient oxygen and energy input, *all* commercial polymeric materials will burn. (Metals also will exhibit some undesirable characteristics under these conditions, i.e., aluminum will melt).

Several methods are available for reducing the fire hazard of polymeric materials. Among these are the following:

1. Development of new polymers whose fire safety characteristics are inherently better than those of the well-known materials; although some materials in this class are available commercially now, most are very expensive.
2. Improvement of the fire safety charcteristics of available, generally lower cost, materials by adding fire retardants; at present this approach is the most important method commercially.
3. Application of a fire resistant coating to the surface of materials or the incorporation of a fire retardant into the bulk at some appropriate stage of processing.

4. Combination of two or more materials in a way that utilizes the best properties of each (e.g., utilizing steel plates on each side of a plywood slab).

Polymers may be fire retarded by introducing into the bulk of the material fillers such as alumina trihydrate or by introducing active compounds referred to as fire retardants. The former lower the fire load either by heat absorption or by providing an inert diluent for the fuel. The latter usually are halogen, phosphorus, nitrogen, antimony, or boron compounds and may be used in synergistic combination. One also distinguishes between reactive and nonreactive retardants according to whether or not they form covalent bonds with the polymer. For more details, see Lyon (1970), Lewin, Atlas and Pearce (1975), and Hindersinn (1977).

The important variables in polymer flammability are summarized in Figure 1. The burning of a polymeric solid is essentially a three-stage process consisting of a heating phase, a thermal pyrolytic phase, and, finally, ignition. The behavior of a polymer during the initial or primary heating phase depends considerably upon its composition. Thermoplastic compositions will generally melt between 100 and 250°C. The loss of rigidity that occurs at the softening point of such thermoplastic materials and the subsequent decrease in melt viscosity as the temperature increases allow these "liquids" to recede from the ignition source rapidly enough in many cases to prevent their subsequent pyrolysis and ignition. In the past, this phenomenon has apparently led to some erroneous conclusions concerning the flammability of such composition (Gouinlock, Porter and Hindersinn 1971). Thermosetting and most natural polymers, such as wood and cellulose, remain essentially unchanged during this early heating stage.

At some later stage in the heating process, thermal decomposition occurs with the release of gaseous products whose flammability will depend upon the chemical composition of the original samples. The temperature and rate at which this stage occurs depends upon the thermal stability of the material and the chemical decomposition reactions occurring under the existing fire conditions. The flammability of a solid is largely determined by its behavior at this stage in the burning process. The establishment of a self-sustaining flame is predominantly dependent upon the generation of sufficient fuel gases from thermal pyrolysis to produce a flammable oxygen-fuel mixture close enough to the solid fuel so that sufficient heat can be transferred from the flame to the solid surface by radiation and/or convection to sustain pyrolysis. This means that the flame zone is usually spatially removed by some small distance from the fuel surface. A separation of flame and solid fuel is necessary in order to allow dilution of the pyrolytic fuel gases with sufficient oxygen to bring the mixture within flammability limits.

Pyrolysis generally proceeds in three closely related phases. In the temperature range of 100 to 250°C, sufficient thermal energy is only available for such low energy reactions as functional group elimination, usually from the end of the chain, and the elimination of such small molecules as water and hydrogen halide. In the 250 to 500°C range, sufficient energy becomes available to break the highest energy chemical bonds usually contained in the structure of most polymers.

These reactions can often lead to the "unzipping" (depolymerization) of polymer

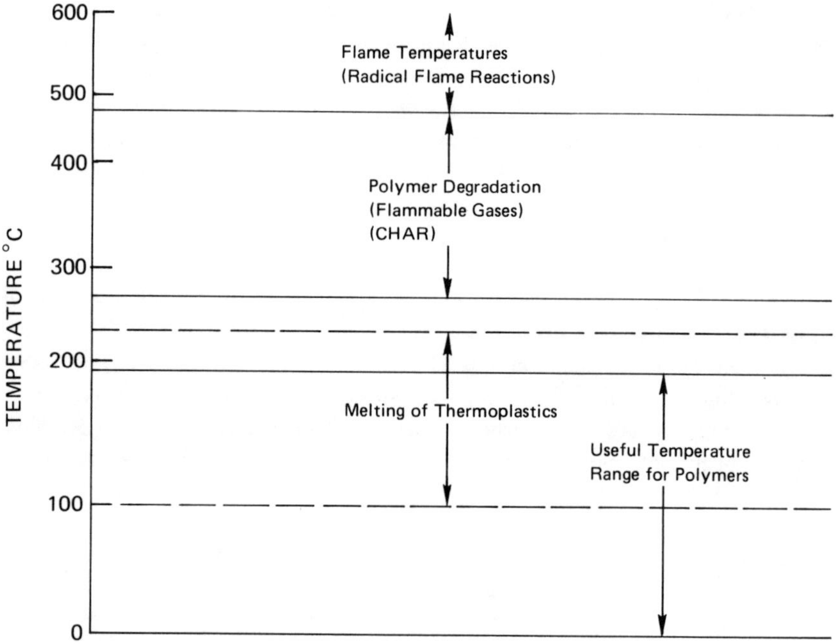

Figure 1. Polymer behavior at various temperatures.

chains that yield flammable monomer or random elimination of small chemical fragments. Both types of products can sustain gas-phase flame reactions. In some cases, however, recombination of some of these fragments also occurs and leads to the formation of aromatic condensed ring systems that are stable under the conditions that pyrolyze aliphatic or alicyclic compounds. In such cases, a third stage of the pyrolysis occurs. Aromatic condensed structures formed in the previous stage are increasingly condensed at temperatures of 500°C, with the eventual elimination of most elements other than carbon. The result is a carbon char, which is highly insulating and difficultly flammable in usual oxygen concentrations. If the char can be maintained in a viscous elastic state during the above intermediate stage, the gases evolved will be trapped in the viscous liquid and, thus, cause the char to expand into a carbonaceous foam. The process for the formation of this special type of pyrolytic char is called *intumescence.* Such char forming reactions are desirable because they convert a flammable polymer to a less flammable char while simultaneously reducing the quantity of flammable gases. If such a conversion can proceed because of the nature of the polymer molecular structure and in the absence of phospohorus, halogen or heavy metal additives, the production of highly toxic by-product gases is minimized and the off-gases are no more toxic than can be attributable to carbon dioxide or carbon monoxide.

At increasingly higher temperatures, the production of gaseous degradation products increases and when a mixture with oxygen in the air is reached that exceeds the flammability limit, ignition occurs. Continued burning at this stage is dependent upon the transfer of sufficient heat from the flame to the condensed phase to maintain an adequate supply of flammable gaseous decomposition of products and upon the supply of oxygen from the surrounding atmosphere sufficient to support combustion. The chemical reactions that generally occur in this flame temperature gas phase are free-radical in nature.

4.1.2.1 Fire Retardant Mechanisms

There are four main mechanisms or methods of altering the flammability of common commercial polymers. The first is the alteration or reduction of the heat of combustion of the total polymer composition. The second is the inhibition of the gas-phase combustion reactions. The third is the alteration of the condensed phase pyrolytic reactions to enhance the formation of char. The fourth is the application of an intumescent coating. Heating of the coating during a fire expands the coating into a thermally stable intumescent char which then protects a substrate from the ignition source. (See Volume 1 for a detailed discussion).

4.1.2.2 Heat of Combustion

It is generally conceded that the incorporation of an organic halogen compound into a polymer, either as an additive or by chemical reaction into polymer structure, will reduce the heat of combustion of the total polymer composition. An example of this effect is indicated in Table 1, where the heat of combustion of various chlorinated methanes is compared to the unchlorinated derivative. It can be seen that successive

Table 1. Heat of Combustion of Hydrocarbons and Their Chlorinated Derivatives.

Substance	Ht. of Combust. (kcal)
CH_4	−181.7
CH_3Cl	−135.1
CH_2Cl_2	− 72.8
$CHCl_3$	− 4.4
CCl_4	+ 47.5

substitutions into the methane series reduce the heat or combustion continuously until the nonflammable carbon tetrachloride exhibits an endothermic heat of combustion. Hindersinn & Wagner (1967) concluded that the heat of combustion was of secondary importance in polymer fire retardance because the fire retardant efficiency in the halogen levels was the reverse of the effect observed in the heat of combustion (i.e., Br > CL > F). As additional support to their conclusion, these authors presented the data summarized in Table 2.

Table 2. Polymer Flammability Ratings and Heat of Combustion Values.

Material	Flammability	Ht. of Combust. kcal/g
Celluloid	Very flammable	− 4.13
Polyethylene	Burns	−11.1
Polystyrene	Burns	− 9.6
Polyethylene (Flame-retarded)	Slow burning	− 9.8
Polyester	Burns	− 4.3
Polyvinyl Chloride	Self-Extinguishing	− 4.3
Polymethyl Methacrylate	Burns	− 6.3
Polmethyl Methacrylate (Flame retarded)	Slow burning	− 5.9

Note: Data from K. Krekeler and Klimke, *Kunstoffe* 1965.

from Table 2 one can see that the highly flammable celluloid (nitro cellulose) has a heat of combustion only slightly more than 4 kcal/g or somewhat less than half of that exhibited by a fire retarded polyethylene composition. Further, the heat of combustion of a slightly flammable polyvinyl chloride is about equivalent to that of the highly flammable celluloid.

4.1.2.3 Gas Phase Inhibition

Gas-phase combustion of hydrocarbon flames has been studied in considerable detail and many of the processes have been quantitatively defined (Krekeler and Klimke, *Kunstoffe* 1965). These reactions have been shown to be predominantly free-radical in nature. The complete oxidation of a hydrocarbon can be explained on the basis of a complicated chain of free-radical reactions. Initial attack is on the saturated hydrocarbons by hydroxyl radicals that are generally the producer of oxygen and hydrogen radicals. In each case, one can see that the products contained highly oxygenated products leading eventually to carbon dioxide and water. Of this series, only reactions of hydroxyl and hydrogen radicals lead to an exponential increase in radical concentration. Inhibition of these reactions thus could disrupt the entire combustion chain.

4.1.2.4 Halogen Inhibition Reactions

The inhibition of branching reactions by hydrogen halide has been suggested by many investigators as a major mechanism for the fire retardant effect of halogens on many polymeric materials. Halogen gas-phase inhibition has been suggested as an explanation for the fact that certain bromocarbons were five to eight times as effective in reducing the flammability of hydrocarbon fuel gas mixtures on a molar basis as such inert agents as carbon dioxide and nitrogen. A mechanism for this flame inhibition has been proposed by Rosser, Wise, and Miller (1959, p. 175) consisting of the replacement of the radical chain carriers in the combustion series previously outlined by less reactive halogen atoms. This suggested mechanism of this type of halogen inhibition is summarized by the following halogen inhibition reactions:

$H\bullet + HX$	$H_2 + X\bullet$	
$OH\bullet + HX$	$H_2O + X\bullet$	Inhibition Reactions
$H\bullet + RH$	$R\bullet + HX$	Regeneration of HX

In this series, the active hydroxyl and hydrogen radicals are converted by reaction with the hydrogen halides into water and the somewhat less active hydrogen molecules. The halogen atoms, which are the other product in this series of reactions, are much less active in oxidation reactions but can react with the other hydrocarbons by extraction of the hydrogen radicals to regenerate the active hydrogen halide catalyst and produce a less reactive carbon radical. Although this mechanism was first proposed to explain the inhibitory effect of halogen compounds upon premixed hydrocarbon flames, its applicability to explain similar inhibitions in polymer diffusion flames is supported by the fact that the order of effectiveness of halogen in polymer fire retardants is the same as that observed in these premixed flames (i.e., $Br > Cl > F$). In addition, it is well established that most if not all of the halogen in the polymer composition is liberated as hydrogen halide on exposure to a flame.

4.1.2.5 Condensed-Phase Reactions

The large body of literature concerning the effectiveness of halogen inhibition of gas-phase reactions has led to its postulation as a primary mechanism for fire retardance of polymers. The gas-phase reactions, however, are only the last in a complicated series of physical processes that must occur in the polymer substrate before an adequate supply of flammable gases becomes available for sustaining an active flame as indicated in Figure 1. Thus, any change in condensed phase reactions leading to a reduced volume of flammable gases could lead to flame extinguishment even in the absence of inhibition of the flame reactions. A careful analysis of the limited available literature concerning the effects of halogens upon condensed phase reactions indicates that alteration of these solid reactions is at least as important in many halogen/polymer fire retardant effects as is the gas-phase inhibition.

The gas phase mechanism is the least attractive form a fire safety point of view because the injection of inhibiting gases into the pyrolytic gas stream leads to incomplete combustion and large volumes of black smoke and most of these inhibitors are toxic to a significant degree and many are highly toxic. It should be noted that the introduction of a fire retardant, while reducing the probability of ignition, may increase the optical density and toxicity of the smoke produced. A condensed-phase mechanism, on the other hand, reduces flammability by conversion to difficultly flammable char with a lesser degree of smoke and generally the production of pyrolytic off-gases no more toxic than carbon monoxide. Char-forming polymer compositions are therefore preferred from a fire safety point of view.

4.1.3 Economic Aspects

Cost is an important aspect of fire hazard reduction. Thermally stable polymers with superior fire safety characteristics may be too expensive for routine use. The application of fire retardant coatings is sometimes a cost-effective approach; however, it is limited to a relatively small number of uses.

The cost of fire retardation by the incorporation of a fire retardant in the polymer varies greatly according to the particular compound or treatment used and to the performance level desired. Fire retardation normally increases the cost of the material except when the desired measure of protection can be obtained with inexpensive inert fillers.

4.1.4 Educational Aspects

During its deliberations, the committee was presented with clear indications that many unnecessary fire hazards result from a general lack of knowledge and appreciation of fire safety problems among the general public in the use of polymeric materials. Many synthetic polymeric materials burn differently from the more familiar natural materials. They may melt and drip and often give off dense and acrid smoke. Many burn with intense flame and resist conventional fire-fighting. These features tend to promote panic in members of the general public and can lead to damage and loss of life that might have been avoided with a better understanding of the material's performance in a fire.

Although public ignorance is understandable, it is difficult to continue to accept the lack of understanding of polymer properties in general, and their fire safety aspects in particular, among practicing engineers, architects, designers, and builders. Often professionals who routinely recommend and specify the use of polymers are without even rudimentary training in polymeric materials. Possibly, the general lack of understanding of the nature and properties of polymers results from the fact that polymer science and engineering is only 30 to 40 years old, and availability of training in these relatively new materials in schools has been generally limited, both at the secondary and tertiary level. (See Volumes 1 and 4 for further discussion).

4.1.5 Planning and Coordination of Research Efforts

In the course of its regular activities, the committee heard presentations by qualified groups describing their problems and programs in the area of fire safety aspects of polymeric materials. On the basis of this information, it became apparent to the committee that the planning, coordination and dissemination of these efforts needed improvement.

4.2 Wood and Wood Products

4.2.1 Introduction

Wood and woodbase products have long been used in the United States for manufacturing passenger vehicles of various types. Although the combustibility of wood has limited its application, proper fire safety design, slow charring rate, and good strength retention of large wooden members under fire has secured good acceptance in the United States of wood as a fire safe material. Considerable technical information is available on methods for the treatment of wood to reduce the flammability and initial rate of heat release and to render it self-extinguishing. However, these treatments have not been widely applied in passenger vehicles because of the limited use of wood in these areas. (The fire-retardant treatment of wood is discussed in some detail in Volume 1.)

4.3 Fibers

4.3.1 Introduction

Fibers form the main components of vehicle cushion covers, carpeting and other compartment furnishings and their flammability characteristics are therefore of much concern. Fibers are discussed here under the headings of natural fibers, synthetic fibers, glass fibers, and fibers from thermally stable polymers.

4.3.2 Natural Fibers

4.3.2.1 Cotton

Cotton is extensively used as a base fabric for polyvinyl chloride type coated fabrics. It is essentially cellulose and cellulose is rich in moderately reactive hydroxyl groups and will burn under a wide variety of conditions. In principle, cotton in the form of fiber, yarn, or fabric can be treated with fire retardants in order to reduce its flammability. Fire retardants for cotton have been reviewed by Drake (1966, 1971). Successful and potentially acceptable durable fire retardants for cotton and rayon fabrics are of two general types — metal oxides and organophosphorus compounds.

Fire retardants based on certain metal oxides, especially in combination with halocarbon, have found greatest use in weather resistant textile products. Large quantities of these materials are generally needed to impart sufficient fire resistance to a fabric to enable it to pass a vertical flame test. The unsatisfactory economics combined with the usually poor hand imparted by the halocarbons and loss during laundering has limited the development of laundry-durable finishes of this type.

Organosphosphorus fire retardant compounds are made to penetrate the fiber where they react, polymerize, or copolymerize with an appropriate monomer or, in some systems, with the cellulose. Another way of applying organophosphorus compounds consists in depositing preformed phosphorus-containing polymers on the fibers or fabrics. Subsequently, these are either further polymerized or fused to provide durability.

Tetrakis(hydroxymethyl)phosphonium chloride (THPC) is a fire retardant for medium and heavy weight cotton fabrics. Finishes based on THPC, urea, methylol melamine, and various textile modifiers have been in use since about 1957 and are perhaps the most important methods for reducing the flammability of cotton. Other phosphorus compounds that have been suggested as fire retardants for cotton, but are much less important than THPC, are tris(1-aziridinyl) phosphine oxide (APO)[1], reactive phosphonates

[1]Recently barred from children's sleepwear due to carcinogenicity upon contact (Consumer Product Safety Commission 1977).

such as (N-methylol dialkyl phosphonopropionamide, phosphoric triamides in conjunction with polyfunctional N-methylol compounds, dialkoxy phosphinyl triazines and diamines of alkyl phosphonic acids.

Good fire resistance generally is obtained on cotton fabrics through insolubilization of about 2 to 3 percent phosphorus, preferably in conjunction with nitrogen. Several excellent reviews have discussed the dozens of experimental materials proposed and the knowledge currently available on the mechanism of fire retardation in cellulose (Lyons 1970, Kasem and Rouette 1972).

4.3.2.2 Wool and Other Natural Fibers

Wool textiles are generally less flammable than cellulosics and are extensively used in carpeting and seat covers. High concentrations of hydrogen cyanide have been found in the pyrolytic off-gases from wool products. The flammability of wool has been reviewed in several recent articles (Benisek 1972, 1972a and 1973. Friedman et al. 1973). The fire retardant methods have not been as extensively studied as those for cotton. The flammability of wool generally is decreased by treatment with organophosphorus compounds or by treatment with specific salts of polyvalent metals.

4.3.3 Synthetic Fibers

Although synthetic fibers represent the "commodity" items of the textile industry and are produced in larger total volumes than cotton and wool combined, they have relatively little utility in transportation vehicles at the present time. Relatively small quantities are used in seat and wall coverings and carpeting. Such fibers include rayon, acetate, nylon, polyester, olefin, acrylic, glass, and others. With exception of glass, no fibers in this group can be considered to offer protection against direct exposure to flame. Despite their negligible use in transportation at present, the continually improving economics of the synthetic fibers could lead to their increased use in the future. Because of this favorable outlook, the following abbreviated discussion of some of the more important synthetics is included for completeness. (See Volume 1, Chapter 3, for more detailed information on this subject.)

4.3.3.1 Fire Safety Aspects of Fiber Blends

Blend fabrics made of yarns containing two or more fibers of different chemical composition and properties have attained great commercial importance in textile markets. Fiber blends pose synergistic fire hazards that often lead to unexpected flammability characteristics (e.g., the inclusion of even small amounts of cotton in a polyester fiber garment can lead to a severe flammability hazard because the nonfusible cotton prevents dripping of the fusible polyester, thus increasing the fuel available for burning). Early investigations have established that the fire safety aspects of blends cannot be predicted from a knowledge of the behavior of individual fiber components (Tesoro and Meiser 1970). The physical and chemical interactions of different fibers in blends under conditions of burning pose complex problems that are not understood and have not been adequately studied. However, a considerable amount of empirical work has been carried out during the past few years on the flammability of blends. Although fabrics have been developed from a wide variety of fiber blends, both synthetic and natural, polyester cellulose types have attained the greatest commercial utility.

4.3.3.2 Polyester/Cellulose Blends

Tesoro (1973) has reviewed the state of the art on the problems of fire retardance in polyester/cellulose blends. Although satisfactory fire retarded products are not available commercially, a research program on this particular problem has been initiated (July 1974) under the auspices of the National Bureau of Standards (NBS) as part of the Experimental Technology Incentives Program.

4.3.3.3 Inorganic Fibers

The fibers discussed in Section 4.3.3 are made from organic polymers. For some end uses, inorganic fibers are important. Glass fibers, for example, melt at about 515°C but do not burn. They are generally treated with organic finishes to enhance their resistance to abrasion and to improve other functional properties. The flammability hazard of glass fibers in actual use is thus significantly modified by the presence of these organic materials. Although glass fibers prepared in this manner are widely used for draperies, curtains, and other similar applications where nonflammability is necessary, they find their greatest utility as reinforcements with many different thermoplastic and thermoset resins. The resulting composites are used extensively in many large-scale applications. Such composites are widely used in transportation vehicles as wall panels, ceiling panels, ventilation ducts, passenger seats, etc. Since the flammability of such products is largely determined by the particular binder used in the composite, this subject is covered below in the sections devoted to the individual polymer or resin system.

4.4 Elastomers

4.4.1 Introduction

The elastomers that account for the bulk of the rubbers used today are shown in Table 3 together with their approximate consumption in the United States in 1972.

Most elastomers burn easily when not fire retarded. Today, there is no elastomer that has the desired combination of low flammability, low smoke emission, good mechanical properties, and reasonable cost (Einhorn 1971; Fabris and Sommer, 1973).

4.4.2 Fire Safety Aspects of Elastomers

Since the fire retardation of natural and synthetic rubbers has been surveyed in detail by Fabris and Sommer (1973), this section is confined to a few general remarks.

The incorporation of halogens, either in an additive or as an integral part of the molecule, has been a prime approach to decreasing the flammability of elastomers. Thus, we have polychloroprene, chlorinated polyolefins, epichlorohydrin rubbers, the various fluoro- and chlorofluoro elastomers, and halogen-containing polyurethanes as

Table 3. 1972 Consumption of New Rubber (million pounds) in the United States.

Type	Consumption
Styrene-Butadiene	3,189
Natural	1,411
Polybutadiene	682
Polychloroprene	273
Butyl[1]	268
Polyisoprene	250
Ethylene-Propylene-Diene	141
Nitrile[2]	141
Chlorosulfonated Polyethylene	50
Polysulfide	50
Silicone	50
Fluoroelastomers	50
Polyacrylates	50
Polybutene	50
Polyisobutylene	50

[1]Includes Chlorobutyl
[2]Includes Nitrile/PVC
Note: Data from Pariser et al. 1974.

well as various compositions in which halogenated additives are used. All these materials are deficient in their fire safety characteristics (Gross et al. 1969); they give off smoke and hydrogen halides on combustion and/or exposure to an intense fire environment. In addition, some rubbers (particularly fluorinated materials) have the potential for generating other specific toxic combustion products.

Phosphorus compounds also have been used to decrease flammability. In elastomers their use is generally limited to plasticized poly(vinyl chloride) and to polyurethanes. Materials fire retarded with phosphorus do have somewhat lower flame propagation and are more difficult to ignite by small ignition sources but show little improvement when exposed to intense fire situations (Trexler 1973).

The third approach to reducing flammability (and smoke) consists of replacing all or part of the carbon in the polymer structure with inorganic elements (Laur 1970). The prime example of this effort is the family of silicone elastomers (Laur 1970, Hooker Chemical Corp. 1970, Pepe 1970) and the developmental phosphonitrilic elastomers (Hagnauer and Schneider 1972).

A fourth approach consists of incorporating large amounts of inorganic fillers. This addition reduces the fuel value of the composition even if the filler has no specific fire retardant properties. Fortunately, most elastomers can tolerate or even require a substantial amount (about 50 percent) of particulate inorganic filler (Trexler 1973).

4.4.3 Specific Elastomers

The fire safety aspects of elastomers are largely determined by their chemical structure. From this point of view they may conveniently be assigned to several distinct groups.

4.4.3.1 Hydrocarbon-Based Elastomers

This group includes natural rubber, synthetic cispolyisoprene, polybutadiene, styrene-butadiene rubber, butyl rubber, and ethylene-propylene rubber as its main constituents (Kennedy and Tornquist 1968, Bateman 1963, Morton 1973, Stern 1967, Winspear 1972, and Witby et al. 1954).

These rubbers are low cost materials with good mechanical properties and, thus, are used in large volume applications such as automobile and truck tires. They do, however, burn readily with production of much smoke. Fire retardant additives reduce flame spread and ease of ignition from low energy ignition sources but do not prevent burning in an intense fire situation. Alumina trihydrate is receiving intensive study as a filler to reduce flammability and smoke in these elastomers (Texas-U.S. Chemical Co. 1964, Hecker 1968, Dalzell and Nulph 1970, Hooker Chemical Co. 1970, Polsar 1970).

4.4.3.2 Chlorine-Containing Elastomers

These elastomers comprise polychloroprene, rubber hydrochloride, chlorinated ethylene polymers and copolymers (chlorinated polyolefins) and epichlorohydrin rubbers (Morton 1973, Stern 1967, Winspear 1972, and Witby et al. 1954). These materials have significantly better fire retardance than the straight hydrocarbon rubber. They, however, generate extensive black smoke and hydrogen chloride gas when exposed to a fully developed fire.

Chlorinated elastomers, particularly polychloroprene (more commonly designated neoprene), are widely used where fire retardance is important. Some large-scale applications in transportation vehicles are in electrical insulation, foam seat cushions, and seat liners over other foam cushioning. At this writing, most new mass transportation vehicles and vehicles undergoing refurbishment employ neoprene foam for seat cushioning where cushioning is employed.

4.4.3.3 Nitrile Elastomers

Nitrile rubbers are copolymers of butadiene (see above) and acrylonitrile (Hofman 1963). The ratio of butadiene to acrylonitrile is similar to the ratio of butadiene to styrene in SBR. The cyanide group imparts to these elastomers some of the properties of the halogen-containing rubbers but also constitutes a potential toxicological hazard.

4.4.3.4 Polyurethane Elastomers

Polyurethanes are polymers containing the group -NH-CO-O- (Saunders and Frisch 1962). They are formed typically (see sections 6.3.2 and 6.3.4) through the reaction of a diisocyanate and a glycol. Because a variety of glycols or esters can be coupled with different diisocyanates, a large number of linear polymers can be obtained in this way. These elastomers are crosslinked by including a controlled amount of a polyfunctional monomer (e.g., a triisocyanate or trihydric alcohol) in the reaction.

Fire retardant grades, generally based on bromine and/or phosphorus containing additives, are available. These, however, still burn in intense fires. Smoke generation is generally less than with hydrocarbon elastomers, but some hydrogen cyanide gas can be generated. The major use of polyurethane elastomers is in foams (seat cushions, insulation, etc.); these are discussed in Section 4.6.

More recently, reaction injection molded (RIM) polyurethanes are finding increasing use on automobile exteriors in such applications as front and rear fender extensions, front fender skirts, and panels and front or rear fender fillers.

4.4.3.5 Polysulfide Elastomers

These rubbers, also known as thiokols, are polymers composed of aliphatic hydrocarbon chains connected by di-, tri-, and tetrasulfide links (Morton 1973, Stern 1967, Whitby et al. 1954). Because of their outstanding resistance to hydrocarbon solvents, they are used extensively in sealants for aircraft fuel tanks and pressurized cabins but have only a very limited utility in land vehicles.

4.4.3.6 Fluorocarbon Elastomers

As a class, fluorocarbons (especially the experimental specialty rubbers) are expensive and are commonly deficient in their mechanical properties for many applications. They are generally difficult to ignite and not prone to propagate flames. They do, however, have potential toxicity hazards from products of combustion and/or pyrolysis in intense fires.

4.4.3.7 Silicone Elastomers

Silicone elastomers generate relatively little smoke, are reasonably fire retardant in air, and, when burned, have low heating values (Dow-Corning Co. 1969, Compton 1967, Karstedt 1970). They burn slowly and produce no flaming drip. They are relatively expensive (less so than the fluorocarbons but more so than the hydrocarbons). For many applications their mechanical properties are marginal. Silicones do, however, offer one of the more promising combinations of fire safety aspects, useful physical properties, and reasonable cost. They are used in electrical insulation and seat cushions.

4.4.3.8 Phosphonitrilic Elastomers

Phosphonitrilic elastomers represent another example of "inorganic elastomers" (Hagnauer and Schneider 1972). The phosphorus-nitrogen backbone:

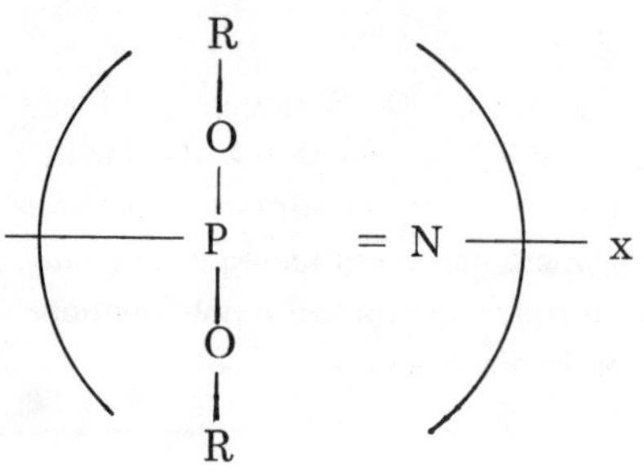

or

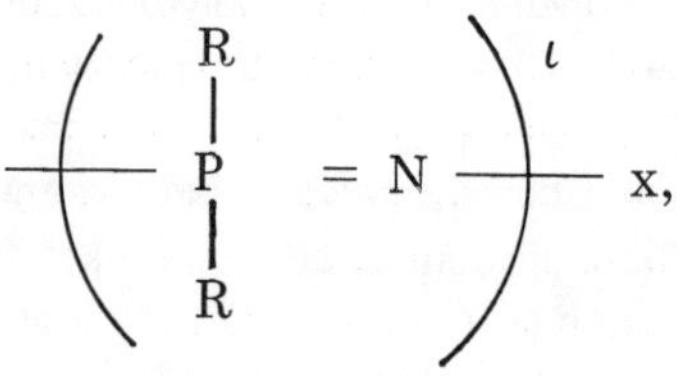

supplies the flexibility required for elastomeric properties and contributes little fuel value. The various side groups (R or R′) affect many of the characteristics of the elastomers including their flammability. For example, long hydrocarbon side chains would increase flammability while fluorocarbon side chains would not, but could contribute to undesirable pyrolysis products.

These phosphonitrilic materials are in the early stages of development and much needs to be done to define their utility and feasibility for various uses. This definition includes data on the combustion and pyrolysis products contributed by the phosphorus and nitrogen. The phosphonitrilics are, however, one of the main hopes for a low smoke, low flammability elastomer. Continued successful development of these new polymers will most likely result in their future extensive use in the transportation industry. Two likely large-scale applications in land passenger vehicles are as seat cushioning and electrical insulation.

4.5 Thermoplastic Resins

4.5.1 Polyolefins

The principally used polyolefins are low density polyethylene, high density polyethylene, polypropylene, and ethylene/propylene copolymers. Smaller volume materials are polybutene, poly 4-methylpentene, ethylene/vinyl acetate, and other copolymers and blends.

It was predicted that more than 160,000 metric tons of polyolefinic plastics would be consumed by transportation applications in 1976 (Modern Plastics 1976). Most of this usage is polypropylene, used extensively in automobiles. Approximately 80 percent of the polyolefinics are used in such automotive applications as battery cases, fan shrouds, and fender liners. More recent applications include inner wheel housings, fender aprons, and timing belt covers.

The combustion of polyolefins has been extensively reviewed by Cullis (1971). Chemically, polyolefins are very similar to paraffin wax. They ignite easily, burn with a smoky (but less smoky than polystyrene) flame and melt as they burn. The mechanism of burning of polyolefins is similar to that for most solid materials. The products of combustion of polyolefins generally are those expected from burning hydrocarbons; the principally toxic material is carbon monoxide (Ball 1973).

Use of additive systems based on the combination of halogen compounds and antimony oxide has been very effective in reducing flammability. Despite greatly improved resistance to ignition under low thermal energy environments can be obtained, all of these compositions burn readily in a fully developed fire, contributing to the fuel load and giving very hot fires.

Although polyolefins present little fire hazard in their present application areas, there is danger of producing excessive fuel loads and introducing paths for rapid flame spread as their use expands, especially when applied as structural foams.

4.5.2 Sytrene Polymers

ABS (Acrylonitrile-butadiene-styrene copolymer), one of the major types of styrene polymers, is finding fairly widespread use in automobiles and trucks as rear lamp housings, grilles (chromium plated), rear seat filler panels, rear window louvre inserts, and headlamp doors (chromium plated).

With the increasing and diverse use of foam (see Section 4.6), and large thermoformed items (e.g., in seats and car furnishings), relative hazard definition by use-analysis and meaningful testing are required. Potential hazards arise when relatively large amounts of polymers are used and when large surface areas are exposed. These hazards are increased by the rapid burning rate of polystyrene, and by the high temperatures and dense smoke generated in polystyrene fires.

A large number of additives has been proposed and tried with a view to reducing the flammability of styrene-based products (Lindemann 1973, Horwath 1973). The majority of these additives are halogen compounds and usually incorporate a synergist such as antimony oxide. Generally, they are believed to function by increasing the depolymerization rate and promoting the dripping of molten polymer, thus removing heat and flame from the burning sample. Their use has been extensive in styrene-based foams (see Section 4.6.5).

4.5.3 Polyvinyl Chloride (PVC)

Vinyl chloride is an inexpensive monomer that can be polymerized by a variety of free radical catalysts to yield a high molecular weight polymer with the general structure:

$$-(CH_2 - \underset{\textstyle Cl}{CH}-)_x$$

PVC itself does not burn under most normal conditions. The poor thermal stability of the polymer, however, generally necessitates compounding with significant and often large amounts of plasticizers or processing aids. Many of these compounding ingredients are flammable, particularly the widely used phthalate, sebacate, and adipate esters and various low molecular weight adipate polyesters.

With the exception of polyurethanes, PVC is the largest volume polymer currently being used in transportation markets today (Modern Plastics 1976). In 1975 approximately 125,000 metric tons was used in passenger automobiles alone as upholstery materials, seat covering, head liner, and in extrusions. Additionally, 10,000 metric tons are being used for similar applications in trucks, buses, rail transport vehicles, and aircraft. Much of the upholstery used in these applications are PVC coated fabrics of various types; cotton is the most commonly used base fabric. Other base fabrics, such as nylon, also are used in PVC composites.

When exposed to a flame or excessive heat, PVC emits hydrogen chloride at relatively low temperatures in a highly endothermic process. This characteristic, together with the fact that the polymer conains more than 50 percent chlorine by weight, accounts for the low flammability of the uncompounded polymer. Depending upon the amount or type of the compounding ingredients used during fabrication, decomposition products may include benzene, hydrocarbons, char, and other fragments.

Chlorinated or phosphorus based plasticizers also are used in large quantities to reduce the flammability of plasticized compositions. Among these, phosphates and chlorinated paraffins are used most widely. Phosphates, particularly tricresyl phosphate, cresyl diphenyl phosphate, and 2-ethylhexyl diphenyl phosphate, have traditionally been added to PVC as plasticizers. They also enhance fire retardance and achieve excellent flame-out times.

4.5.4. Acrylics

Polyacrylates are extensively used in the automotive and transportation markets primarily as decorative paints and coatings. Significant quantities of polymethylmethacrylate are also used as glazing because of good transmissivity, relatively high impact resistance and resistance to light degradation. PVC modified acrylic panels are also used as decorative wall and ceiling finishes in some rail passenger cars. Approximately 18,000 metric tons of acrylic polymers was used for these purposes in 1976, largely in passenger automobiles, trucks and buses (Modern Plastics 1976).

The acrylics are polymers formed from the acrylic (R = H) or methacrylic (R = CH_3) esters according to the formula:

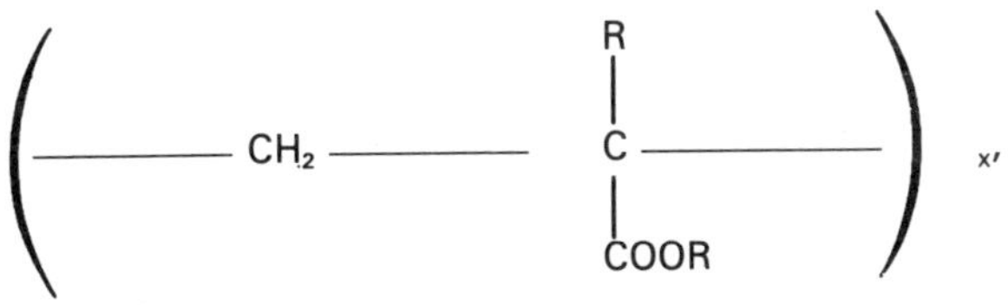

where R′ represents an alkyl radical. Polyacrylate elastomers have been discussed in Section 4.3.4. The major plastic in this group is the homopolymer of methyl methacrylate R′ = CH_3), a crystal clear material that softens at about 100°C.

Poly(methyl methacrylate), (PMMA), ignites readily and softens as it burns. Burning rate, fuel load, and smoke production are less than those for polystyrene (Hilado 1969). In burning, PMMA undergoes depolymerization to the monomer from the heat of the ignition source, the heat of combustion, or other environmental energy (Conley 1970). The volatile products of pyrolysis then burn in the gas phase.

Halogen and antimony compounds have been used to reduce burning rates and ease of ignition (Howarth 1973). Less effort has been devoted to the fire retardation of PMMA than to that of other polymers due partly to the realization that it is difficult to retard the "unizpping" depolymerization mechanism so characteristic of this polymer for most applications. Moreover, fire retarding additives usually detract from the excellent transparency and aging characteristics of the polymer.

When PMMA is used as glazing, particularly in relatively large areas, the potential fire hazard should be carefully analyzed. The use of acrylic glazing in urban transit vehicles has led to an increased fire hazard owing to the ready combustion of the windows.

4.5.5 Nylons

Recent applications for mineral filled nylon in automobiles include rear fender extensions, headlamp doors, fender louvres, and filler caps for fuel tanks.

The nylons currently are used in relatively small items and these have not, in themselves, posed serious fire safety problems. However, as larger items (e.g., gas tanks, large castings, automotive body parts, and other structural applications) are considered, more attention needs to be given to the analysis of fire hazards that might be introduced. Nylon carpeting also is used in some rail passenger cars. Modern Plastics (1976) estimated about 14,000 metric tons of nylon would be used in 1976 in various transportation markets.

The major nylon plastics are nylon 66:

$$\text{+}NH - (CH_2)_6 - NH-CO\,(CH_2)_4CO\text{+}_x$$

and Nylon 6:

$$\text{+}NH - (CH_2)_5\,CO\text{+}_x$$

Nylon moldings often are described as "self-extinguishing." This effect is due largely to their tendency to drip when ignited. Dripping removes the flame front and hot polymer from the burning piece. If dripping is prevented, nylons burn with a smoky flame. Nylons pyrolyze to a complex mixture of hydrocarbons, cyclic ketones, esters, and nitriles with some carbonization (Conley and Gaudiana 1970). The volatiles burn to a variety of products depending on conditions.

Fire retarded formulations are generally based on phosphorus or on halogen-containing additives with or without the addition of antimony or iron oxides (Howarth 1973, Pearce 1975). Hydrated alumina also has been used as a fire retardant. None of these systems prevents burning in a fully developed fire.

4.5.6 Cellulosics

More than 16,000 metric tons of cellulosics were forecast to be used in transportation in 1976 (Modern Plastics 1976). These materials are used largely in paints and lacquers. Cellulose esters also are used as the organic binder in safety glass.

Cellulose derivatives, useful as plastics, include regenerated cellulose, organic and inorganic esters of cellulose, and cellulose ethers. Their basic building block is the substituted glucose unit shown below:

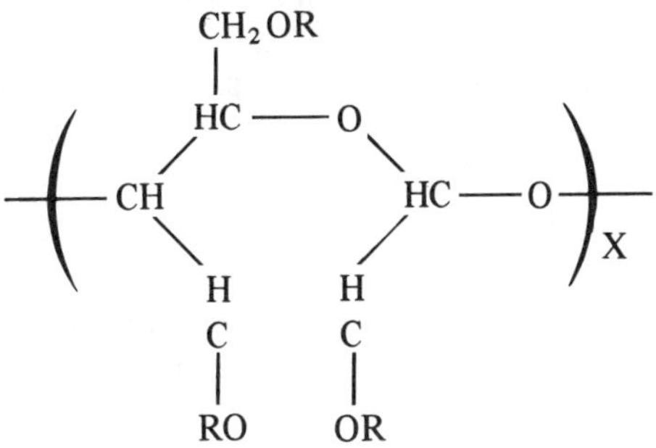

The formula for cellulose is obtained when R = H.
Organic cellulose esters and ethers burn with a yellow sooty smoke. They melt and drip as they burn. Halogen and phosphorus-containing plasticizers have been used to give fire retardant grades (Howarth 1973) with reduced ease of ignition and flame spread, but, these still burn readily in a fully developed fire.

4.5.7 Polyacetals

Little success has been achieved in making polyacetals more fire retardant because of the mode of their pyrolysis and the nature of their pyrolysis products. On the other hand, many of these polymers are used in relatively small parts in vehicles where they do not present major fire hazards.

Commercial polyacetals are formaldehyde polymers and copolymers terminated (capped) with ester or other groups for stabilization. The simplest polyacetal is poly(methylene oxide):

$$-\!\!\left(CH_2 \text{———} O\right)\!\!-_x$$

Three principle flammability characteristics of formaldehyde polymers and copolymers are low oxygen requirement for combustion (low LOI), very little smoke, and low fuel value. These characteristics result from the unique chemical structure. Formaldehyde polymers depolymerize to monomer at relatively low temperatures (230°C) (Conley and Gaudiana 1970). The evolved formaldehyde burns with a clear blue flame accompanied by some polymer melting and dripping, depending on molecular weight, fillers, geometry, etc. Since formaldehyde is oxygenated and has no carbon-carbon bonds, it burns without smoke even at low oxygen levels. Its products of combustion are water, carbon dioxide, and some carbon monoxide.

4.5.8 Polyesters

Polyesters are increasingly used in components of transport vehicles. A recent example is replacement of zinc die cast headlamp moldings by polyesters.

The polyesters considered here are the linear thermoplastics; poly(ethylene terephthalate), (PET), poly(tetramethylene terephthalate), (PTMT), and their modifications. Crosslinked styrenated polyesters are discussed among the thermosetting materials in Section 4.5.4.2.

These polymers burn with a smoky flame accompanied by melting and dripping and little char formation. Fire retarded grades are generally prepared by incorporating halogen-containing materials as part of the polymer molecule or as additives. Metal oxide synergists are frequently included. These fire retarded systems are resistant to small ignition sources in low heat flux environments, but they burn readily in fully developed fires.

4.5.9 Polycarbonates

Polycarbonates are a special class of polyesters derived from bisphenols and phosgene. The commercial products are based on bisphenol-A.

Almost 5,000 metric tons of polycarbonate was predicted for use in transportation application during 1976 (Modern Plastics 1976) with more than half that amount being used in other than automotive applications. Perhaps its largest use is as impact resistant windows in buses, rail transport, and aircraft. Polycarbonate also is used in a variety of impact resistant moldings for automobiles, trucks, and buses.

Commercial unmodified polycarbonates are significantly less flammable than unmodified styrene, olefin, or acrylic polymers. On pyrolysis or burning they produce some char. They extinguish during simple horizontal burning tests and have an oxygen index (Hilado 1969) significantly above those of all the previously discussed unmodified thermoplastics (see Table 9, Volume 1). Their fire resistance has been further improved by the use of halogenated bisphenols in the preparation of the polymer or by the use of halogen-containing additives with or without antimony oxide (Howarth 1973). Recently a number of patents have issued which describe the use of small amounts of perfluoroalkane and aryl sulfonates as excellent flame retardants for polycarbonates in the absence of halogenated compounds (Nouvertne 1973, Mark 1975). Polycarbonates will burn under high thermal fluxes. Little information is available on the toxicity of their combustion products, but those containing halogen would be expected to evolve some hydrogen halide.

4.5.10 Chlorinated and Chlorosulfonated Polyethylene

Chlorinated and chlorosulfonated polyethylene are relatively low volume polymers in the transportation industry. The latter is used extensively as fire resistant electrical wire and cable insulation in rail passenger cars where its flammability has created some fire safety problems.

Polyethylene can be chlorinated in the presence of light or free radical catalyst to give a chlorinated polymer of the general structure (Canterino 1967):

$$\left[(CH_2 - CH_2)_x - (CH_2 - \underset{\displaystyle Cl}{\underset{|}{CH}})_y - (CH_2CCl_2)_z \right]_n$$

The chlorine content, and therefore the properties of the product, varies considerably depending upon the extent of chlorination and the reaction conditions. The flammability decreases directly with the increase in chlorine content. Various compositions are reported to extinguish under ASTM C 635 conditions. These are reported to contain 25 to 40 percent chlorine by weight. Compositions containing as much as 67 percent chlorine have been prepared. As with chlorinated polymers in general, antimony oxide enhances the efficiency of the halogen and reduces the amount of chlorine required to yield the desired fire retardant properties.

The flammability characteristics of these materials resemble those of poly(vinyl chloride) and poly(vinylidene chloride). Hydrogen chloride is a major combustion product.

Polyethylene can be chlorosulfonated by methods similar to those used in the chlorination already described to give polymers of the general structure:

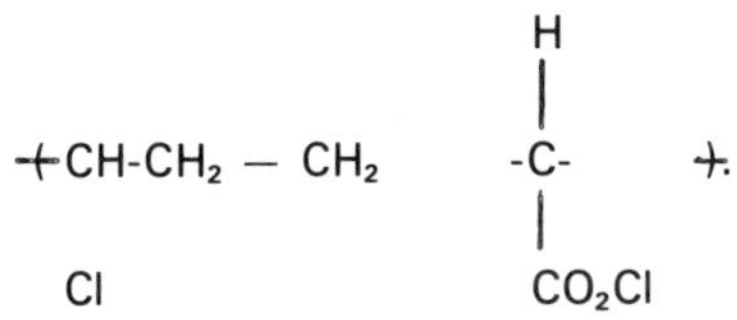

The properties of this composition can be varied widely depending upon the extent of chlorosulfonation.

As expected, the flammability of the polyethylene is reduced as the chlorosulfonyl chloride content is increased. Its combustion has received little study; hydrogen chloride and sulfur dioxide are presumed to be major products of combustion.

4.6 Thermosetting Resins

Thermosetting resins are produced in large quantities. They are used in seat frames, wall panels, and other vehicle components where they can contribute significantly to the fire load. Consequently, their fire safety characteristics are of primary concern.

Thermoset polymers are distinguished from the thermoplastics discussed in Section 4.5.3 in that they become chemically crosslinked during final molding. The final product is then set into shape by primary chemical bonds, and for most practical purposes no longer can be melted, reshaped, or dissolved.

Because of their brittle nature, thermosets are used almost exclusively in conjunction with various inorganic or organic fibrous reinforcements and various types of powdered fillers. In many cases, these reinforcements and fillers comprise more than half of the final composition and can alter the flammability or fire safety of the total composite significantly. It is important, therefore, to consider the total composition before deciding upon the flammability characteristics or fire safety aspects of these materials.

Because of their crosslinked nature, thermosets generally do not soften or drip when exposed to a flame as do many thermoplastic materials. Their flammability is a function of the thermal stability of the primary chemical bonds and the ease with which volatile gaseous products can be produced by pyrolytic processes to provide fuel for a self-sustaining fire. Many thermosets (e.g., the phenolic resins) produce very little flammable fuel when heated by an ignition source but degrade into a difficultly flammable and insulating char that can only be oxidized at extremely high temperatures and/or high oxygen concentrations. Burning of such materials can be a slow process under many conditions since the polymer substrate is protected by the surface char. These resins are inherently fire retardant and will pass many common laboratory tests without the need of a fire retardant modification or additive. Their fire retardance, however, is a function of the mechanical stability of the insulating char and is limited by the resistance of elemental carbon to oxidation.

4.6.1 Phenolic Resins and Molding Compounds

Phenolic resins of various types find considerable utility in transportation applications but mostly in small-scale moldings where flammability is not a factor. More than 25,000 metric tons of phenolics were predicted to be used for this purpose in 1976 (Modern Plastics 1976). Representative applications are as distributor caps, battery cases, transmission gears, and pulleys. Phenolic adhesives also are used extensively in plywood manufacture and, thus, are a minor constituent of plymetal sandwich composites used in rail car floor panels.

Two general types of first stage phenolic resins are produced depending on the catalyst, the phenol/formaldehyde ratio, and the reaction conditions. These resins are called resoles and novolaks, respectively. The physical properties of phenolic resins vary widely depending upon the type, kind, and amount of filler; kind of reinforcement used; phenol to formaldehyde ratio; type of curing catalyst; and other formulation variables. More information may be found in Brydson (1975), Billmeyer (1971), Foy (1969).

4.6.1.1 Resoles

These liquid resinous products are prepared by the condensation of phenol with formaldehyde in ratios from 1:1 to as high as 1:3 in the presence of an alkaline catalyst. The resole is normally set (or cured) by a simple thermal treatment although room temperature cures are possible by the addition of a suitably active catalyst.

4.6.1.2 Novolaks

If a phenol to formaldehyde ratio of 1:1 or 1:less than 1 is used in the initial condensation with an acidic catalyst, thermally stable resinous thermoplastic polymers, commonly referred to as novolaks, can be formed. These resins can be formulated subsequently with fillers, colorants, reinforcing agents, catalysts, and additional formaldehyde (or formaldehyde generator) for the form the commercially important molding compounds. The most common formaldehyde substitute used in these resins is hexamethylene tetramine.

4.6.1.3 Fire Safety Characteristics of Cured Phenolics

Cured phenolic resins do not ignite easily because of their high thermal stability and high charring tendency in the presence of fire and heat (Sunshine 1973). The flammability of the end products can vary widely, as mentioned above, depending upon the amount and type of filler used, the crosslink density, the amount and type of reinforcement, and other less important formulation variables. The principal volatile decomposition products are methane, acetone, carbon monoxide, propanol, and propane. A variety of additives has been found to be useful in applications where a degree of fire retardance above that inherent in the polymer is required.

4.6.2 Unsaturated Polyester Resins

Because of their versatility and low cost, glass reinforced polyester panels are finding increasing usage as automobile body parts and grille opening panels. They also are being used as body ends, ceiling liners, and window masks in the newer passenger rail vehicles. Modern Plastics (1976) estimated that more than 100,000 metric tons of polyester resin would be consumed by the transportation markets in 1976 with 88,000 tons being used in passenger automobiles. Fire retardant compositions were only a small part (approximately 5.5 percent in 1973) of the total polyester production of 550,000 metric tons in 1974 (Nametz 1967, Roberts et al. 1964).

Two classes of thermoset resins are commonly referred to as polyester resins. These are the alkyds (see Section 4.5.4.6) and the so-called unsaturated polyester resins. The latter are prepared by condensing a saturated dibasic alcohol and both a saturated and unsaturated dicarboxylic acid into a prepolymer (or first stage) resin. The prepolymer is then dissolved in a vinyl monomer, usually styrene. The cured resin is produced by free-radical copolymerization of the styrene monomers and the unsaturated acid residues. The resins are usually compounded with a reinforcing fabric (generally glass cloth or mat) and/or filler before curing.

The burning characteristics of unsaturated polyesters can be modified by the addition of inorganic fillers; the addition of organic fire retardants; the chemical modification of the acid, alcohol, or unsaturated monomer component; and the chemical combination of organometallic compounds with the resin.

A wide variation in flammability characteristics can be achieved in polyester resins by using one or more of the fire retardant modifications described above. Flame spread ratings of 25 or less, as measured by ASTM E84, have been attained by using chlorendic acid with antimony oxide as synergist. These low flame spread ratings can now be obtained in the absence of opacifying antimony using the more efficient bromine monomers.

Both fire retarded and unretarded polyester resin formulations yield copious amounts of smoke when exposed to fire because styrene is the major product of pyrolytic decomposition and styrene burns with a very smoky flame. The high smoke values have only been marginally reduced to date by the use of relatively large amounts of inorganic fillers such as alumina trihydrate.

The relative toxicity of halogenated polyester resins has been a subject of considerable discussion ever since their introduction in 1953. Generally, the chlorine contained in these compositions has been shown to convert largely, if not quantitatively, into hydrogen chloride. Trace amounts of phosgene have been identified as well.

4.6.3 Epoxy Resins

Epoxy resins generally are prepared by reacting a first stage polyfunctional epoxy compound or resin with a basic or acidic crosslinker (or "hardener") to yield a thermoset product crosslinked by ether or ester linkages. The basic epoxy resin can be prepared in a variety of ways although the most common is the reaction of a polyphenolic compound with epichlorohydrin.

The prepolymer can be cured with a variety of crosslinking agents (often incorrectly called catalysts) through the epoxy and hydroxyl groups. These hardeners can be based on amines, anhydrides, and Lewis acids.

Although epoxy resins are normally flammable (Conley and Quinn 1975, Lyons 1970), their flammability can be reduced considerably by the use of a variety of phosphorus or halogen containing additives or reactive monomers. The need for fire retardance in epoxy resins has been relatively small to date and the amount of fire retardant epoxy resins sold yearly is small. The relatively low sales volume of epoxy resins prevents a major fire hazard from these compositions.

4.6.4 Furan Resins

Although furan resins are currently little used in transportation applications, more extensive applications could develop in the future. Furan resins are prepared by reacting furfuryl alcohol and an aldehyde — most frequently formaldehyde (Siegfried 1967). Urea is often used as a modifying agent. The resins are hardened *in situ* with an acidic substance added just before application. A typical curing agent would be p-toluenesulfonic acid.

Although no specific literature reference describes fire retardant methods for furan resins, fire retardant formulations are available commercially.

Furan resins are char-forming and have a low flammability when exposed to a flame, but flame spread ratings (ASTM E84) of 25 or less cannot be obtained in the absence of a suitable fire retardant system.

4.6.5 Amine Resins

The various amine resins are little used in structural parts of transportation vehicles. However, they are finding limited applications in mass transit passenger vehicles as decorative coatings on a variety of metal and reinforced plastic panel substrates. Their high hardness, durability, abrasion resistance, and easy colorability are the prime reasons for their use in these applications.

Amine resins are thermoset resins prepared by the reaction of an amino compound with an aldehyde. The reactive amino groups (-NH_2 or -NH-) are characteristically present as amides. The two most important commercial materials are based on urea and melamine. Although many different aldehydes have been explored as components for amine resins, formaldehyde is the only one of commercial significance (Widmer 1965), Sunshine 1973). In contrast to phenol/formaldehyde resins, amine resins are colorless and generally harder. The technology used in their preparation is very similar to that of phenolic resins.

Little work has been done to develop fire retardance in amine resins because of their relatively high heat resistance, low flammability, and predominant use in applications where flammability is relatively unimportant. A variety of phosphorus and boron compounds have been used to reduce flammability.

4.7 Specialty Plastics

This somewhat arbitrary subdivision comprises those materials that are relatively high priced, have certain particularly outstanding properties, and are produced in relatively small volumes for specialty applications. Until the prices of these polymers are reduced, there is little prospect for their large scale usage in land transport vehicles. Materials in this group fall into two general categories: (1) the aromatic and heteroaromatic polymers that are generally used for their high temperature capabilities, and (2) the fluoropolymers that are generally used for their resistance to temperature, chemicals, and/or combustion. (There is a detailed discussion in Volume 1 of such polymers.)

Polyphenylene oxide (PPO) is one of the polymers in this category that has attained significant utility. It is prepared by the oxidative coupling of 2,6-xylenol to give:

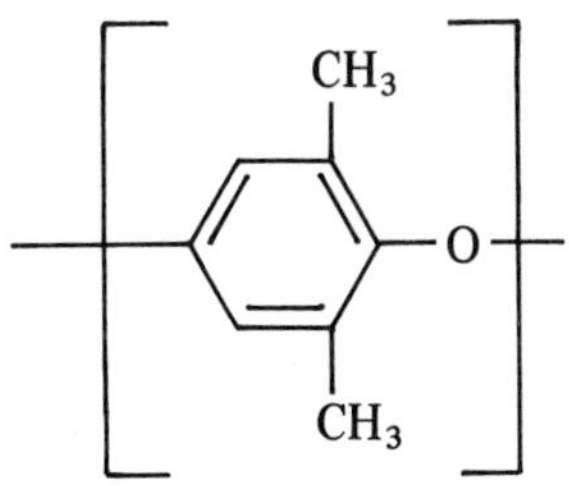

This material is a rigid, tough, chemically and thermally resistant thermoplastic with good electrical properties. It is used for various engineering and electrical applications. Major consumption is in blends with polystyrenes. These blends (trade name Noryl®) have thermal and mechanical properties and costs intermediate between ABS resins and polycarbonates. Glass-filled modifications are available. Major applications in automotive parts are inside the vehicle (i.e., in instrument panel components, speakers and defroster grilles, connectors, and A pillars). Wheel covers represent a growing market on external parts of cars.

Poly(aryl ethers) are char forming and have flammability characteristics similar to those of some of the polysulfones. Blends with polystyrene reduce the dripping of polystyrene and make ignition more difficult. Fire retarded versions, based on halogenated coreactants of organo-phosphate additives are available.

4.8 Foams

4.8.1 Introduction

Foamed polymers are extensively used for seat cushions, padding, insulation, etc. Foamed polymers pose special fire hazards in transport vehicles and their fire safety aspects require special consideration.

Polymeric foams are generally complex multicomponent systems that also may contain fibers and various fillers. They can be divided into rigid and flexible types. In one type, the cellular structure is obtained with the aid of a blowing or foaming agent, which may be either a liquid that vaporizes during the manufacturing processes or a solid that decomposes to give off gas. The gas (CO_2 or water vapor) also may be formed as part of the polymerization reaction. Syntactic foams are essentially polymers which contain tiny hollow spheres of another polymer or glass as filler. Flexible foams generally contain an open cell structure. Rigid foams usually have closed cells.

4.8.2 Fire Safety Aspects of Foams

Since the burning of the polymer combustion occurs only on its surface, the amount of surface area available for combustion is important in determining the rate of combustion and, therefore, the intensity of the flame. Thus, a film burns more readily than a thick molded part. The high surface area per unit weight of foam being subjected to pyrolytic conditions necessarily increases its flammability over that of the solid polymer composition from which it is made. The burning of polymer foams, therefore, differs in several respects from the burning of solid polymers. The most important differences are the lower density and the highly insulative properties of the foamed material.

Since a greater surface is exposed to oxygen in the air, the rate of pyrolysis and burning of a foam is increased relative to that of the base polymer. Low thermal conductivity of foams tends to concentrate the heat on the surface of the structure rather than dissipating it to underlying material or substrate. The result is a rapid heatup and pyrolysis of the surface material when exposed to a flame; this often leads to an extremely rapid flame spread rate. However, other factors may moderate this effect considerably (e.g., the small amount of potentially flammable material per unit volume in low-density foams results in a very small amount of total heat being available per unit area for flame propagation).

Thus, if one considers a thermoplastic such as polystyrene foam, the heat of a flame rapidly melts the foam adjacent to it. The material may recede so fast from the flame front that ignition ceases. If ignition does occur, the fire is extinguished when the flaming liquid drops away carrying the flame front with it.

A highly crosslinked thermoset foam, on the other hand, behaves in an entirely different manner. Since little or no melting occurs, the surface does not recede from the flame front, and the foam is rapidly ignited. The flame then spreads according to the flammability of the foam. A fire retarded foam, under the same conditions, may pyrolyze rapidly in the vicinity of the flame, leaving a carbonaceous char on the surface of the material. This highly insulating char protects the remainder of the material from the effects of the flame. In this manner, a relatively flammable solid is converted into less flammable carbon. Since carbon itself is combustible, the continued impingement of radiant heat flux, accompanied by sufficient oxygen, can generate continued combustion, but the low density of the surface char generally does not allow the retention of sufficient heat to sustain burning in the absence of a sufficient flux of heat to the surface.

4.8.3 Flexible Foams

Flexible foams can be made from practically any elastomer. They are used in a variety of applications, the most important being seat cushioning, carpet underlay, and carpet backsizing. Flexible foams, predominantly polyurethanes, are used extensively as cushioning materials, microcellular bumpers and padding throughout the transportation industry. Recent market estimates (Modern Plastics 1976) indicate that almost 250,000 metric tons will be consumed in transportation with 200,000 tons used in passenger automobiles alone.

When a chemical blowing agent is used in a dry-compounding recipe, the foam rubber is generally referred to as sponge rubber. Sponge rubber is made mostly from natural and from styrene-butadiene rubber, although silicone and fluoro-carbon (Viton) sponge rubbers are also available.

Latex foam rubber is made by beating air into compounded rubber latex. Fluorocarbons are used as (additional) foaming agents in some processes. A gelling agent such as sodium silicofluoride or ammonium acetate may or may not be used (Morton 1973). Again, natural or styrene-butadiene rubber or blends of the two are widely used.

Flexible polyurethane (PU) foams form a high percentage of all elastomeric foams. According to Gmitter and Maxey (1969) cited in Frisch and Saunders (1972), flexible slab polyurethane foam accounts for about two-thirds of all flexible foam. The method of preparation of flexible PU foam is not essentially different from that of rigid PU foam. Flexibility is achieved by appropriately varying the molecular weight and functionality of the polyols used in preparation. Most flexible PU foams today are made from polyethers.

4.8.4 Rigid Foams

Although rigid foams are used extensively in a wide variety of applications, they are relatively little used in transportation vehicles. Among the limited applications for rigid foams in vehicles are thermal insulation, rigid components, and sound barriers. Newer applications include glove compartment doors and passenger seats. Because of their imited application, some of the more common rigid foams such as polystyrene and polyvinyl chloride foams will not be discussed here, and the reader is referred to Volume 1 for details on these materials.

4.8.4.1 Rigid Polyurethane Foams

The relatively small market for rigid foams in transportation is largely for polyurethane. As mentioned above, they are used predominantly as thermal and sonic insulation. More recent applications are automobile floor padding and firewall acoustical padding.

Polyurethanes are the reaction products of a di- or polyhydroxylic compound or resin and a di- or polyisocyanate (Pigott 1969). Polyurethane foams are prepared by a modification of the fundamental reaction of an isocyanate and an alcohol whereby a permanent cellular structure is produced by the controlled introduction of a gas phase in the basic polyurethane during its polymerization. Polyurethane foam technology has been reviewed in detail by Backus and Gemeinhardt (1973) and Gmitter, Fabris, and Maxey (1972).

The cellular nature and low thermal stability of polyurethane foams generally influences their flammability. Because of their low thermal conductivity, the high surface heat flux generated from an ignition source can cause extremely rapid conversion to flammable gases. Such gasification often results in high surface flame spread and high flame temperatures after surface ignition.

In general, a measure of fire retardance is imparted to polyurethane foams by the chemical incorporation of halogens and/or phosphorus compounds. The use of phosphorus in fire retardant polyurethane foams leads to high char formation combined with easy processing because of the relatively low viscosity of most phosphorus compounds. This combination of desirable properties has made phosphorus compounds, with or without halogen, the most widely used fire retardants in polyurethane technology. Reactive phosphorus compounds, such as Fyrol 6 (Stauffer Chemical Company):

$$(CH_3CH_2O)_2\overset{\overset{\displaystyle O}{\|}}{P}-CH_2-N(CH_2CH_2OH)_2\,,$$

are used extensively. They are added directly to the polyol. Polyurethane foams may also be fire retarded by the incorporation of nonreactive additives that act as fillers or plasticizers. The most commonly used example of the latter is tris-(2,3-dibromopropyl)phosphate. This additive has recently come into disfavor because of its reported mutagenic/carcinogenic properties. Nonreactive additives have not been used extensively in polyurethane technology because of their fugitive nature and their tendency to migrate from the foam under many conditions of extended use.

Although polyurethanes themselves are nontoxic, their pyrolytic and/or combustion gases have been shown to contain considerable quantities of toxic gases. Significant amounts of hydrogen cyanide have been detected in polyurethane combustion products (Sumi and Tsuchiya 1973) although the relative toxicity of these materials in gaseous mixtures containing large amounts of carbon monoxide has not been definitely established.

Approaches to fire retardation of flexible polyurethane foams are essentially the same as those used with rigid polyurethane foams. Unfortunately, flexible PU foams burn readily even when fire retarded. A completely satisfactory solution to the serious problem of achieving fire retardant flexible foam for cushioning has not yet been developed.

4.9 Fire Retardant Coatings

4.9.1 Introduction

The use of fire retardant coatings is one of the oldest methods for preventing flammable and nonflammable substrates from reaching ignition or softening temperatures. The two main types of fire retardant coatings are the intumescent and non-intumescent varieties. Fire retardant coatings, particularly, can be used to reduce the flame spread rate of almost any type of organic substrate. Although they are currently little used in transportation, the present public pressure for reduced flammability and smoke generation could lead to their increased use as decorative coatings for the interior of mass transit vehicles. Because of this possibility, a short review of fire retardant coating technology is included for completeness.

4.9.2 Paints and Coatings

Non-intumescent coatings do not provide the same degree of fire protection to the substrate as intumescent coatings; nevertheless, they do not enhance the spread of flame by rapid combustion or contribute a significant amount of fuel to the fire.

4.9.2.1 Alkyd Coatings

The most popular non-intumescent, fire retardant coatings are based on chlorinated alkyds predominantly prepared from chlorendic anhydride (tetrachlorophthalic anhydride). By using suitable chlorinated acids, coatings can be made that have properties comparable to those of conventional coatings in addition to being fire retardant. It is this performance at relatively low cost that has made chlorinated alkyd coatings so successful.

The addition of fire retardant additives to alkyd resins is also commonly employed. Halogenated additives such as chlorinated paraffins have been used and are the most commonly employed additives in these coatings because of their low cost. Antimony oxide is the most commonly used synergist in these applications.

4.9.2.2 Miscellaneous Coatings

Other polymers have been used to a much lesser extent in non-intumescent coatings. Such coatings based upon urethanes (Saunders and Frisch 1962) and epoxy resins (Lyons 1970) have been described. Fire retardant heat cured coatings based upon melamine/formaldehyde and phenol/formaldehye resins may find significant utility as a char resistant coating on factory coated wood.

4.9.3 Conventional Intumescent Coatings

Webster defines intumescence as "an enlarging, swelling or bubbling up (as under the action of heat)." Intumescent coatings are used to protect flammable substrates such as wood and plastics from reaching ignition temperatures. They also protect nonflammable substrates, such as metals, by preventing them from reaching softening or melting temperatures. A thorough review of intumescent coatings was published by Vandersall (1971).

Conventional intumescent coatings contain several key ingredients that are necessary to bring about the intumescent action. An intumescent catalyst is used to trigger the first of several chemical reactions that occur in the coating film. A carbonific compound is included that reacts with the intumescent catalyst to form a carbon residue. A spumific compound is included that decomposes to produce large quantities of gas which cause the carbonaceous char to foam into a protective layer. A resin binder forms a skin over the foam and keeps the trapped gases from escaping. Apart from these key ingredients, intumescent coatings also may include many other constituents used in conventional coatings such as pigments, driers, leveling agents, and thinners.

4.9.4 Nonconventional Intumescent Coatings

By definition, nonconventional intumescent coatings are those in which the elements of intumescence are built into the resin itself. A few such coatings have been recently described; for example, a clear intumescent epoxy coating (Blair et al. 1972) has been prepared by the reaction of triphenyl phosphite with an epoxy resin prepared from epichlorohydrin and bisphenol A. The coating was prepared by adding the amine caalyst to the premixed epoxy-(triphenyl phosphite) resin just before it was applied. The coating consisted of 100 percent solids.

A similar fire retardant clear intumescent urethane coating was reported by Clark et al. (1967, 1968). This coating used a moisture curing polyurethane prepared from pentachlorophenoxy-glycerol ether, triethylene glycol and toluene diisocyanate in a solvent.

4.10 Conclusions and Recommendations

Conclusion: Given sufficient oxygen and thermal energy input, all organic polymers will burn. Billions of pounds of synthetic and natural polymers are used annually in the United States without presenting unusual fire safety problems. However, some uses of polymeric materials in transport vehicles have seriously augmented the fire hazard. *Recommendation:* Support approaches to improve the fire safety of the high volume and low cost polymers.

Conclusion: Many synthetic organic polymers burn in a manner different from that of the more familiar natural polymers such as wood, paper, cotton, or wool. Some synthetics burn much faster, some give off much more smoke, some evolve different noxious and toxic gases, and some melt and drip. Others burn less readily than the natural polymers. *Recommendation:* Initiate a program to define the critical overall fire safety parameters (flammability, smoke and toxic gases, etc.) of polymer based materials. Develop a sound education program to better acquaint the public with the fire safety aspects of polymeric materials and products. Create an overall program to categorize and communicate the goals and results of government-supported work on fire safety of polymeric materials. Initiate programs to determine the relationship of chemical and physical components of polymeric materials to the evolution of smoke and toxic gas formation.

Conclusion: The flammability of many polymers has been improved by the incorporation of hydrated alumina and/or compounds containing halogens, phosphorus, and/or antimony. Many polymeric materials with improved flammability characteristics are deficient in other respects (e.g. costs are high, fabrication is difficult, and combustion products are toxic and corrosive). Blends of cotton and polyester fibers make very desirable and economical fabrics for seat covers, etc.; however, such blends cannot be fire retarded with effective and economical treatments. *Recommendation:* Establish incentives to accelerate the introduction and commercialization of new materials with improved fire safety characteristics. Initiate programs to increase basic knowledge of the relationship between the chemical and physical properties of polymers and fire dynamics parameters and the way this relationship is affected by aging.

Conclusion: Intumescent coatings can be used to enhance the fire safety of some polymeric products. *Recommendation:* Expand materials and application studies on intumescent coatings with emphasis on lowering cost and improving coating performance.

Conclusion: Designers and engineers of land transport vehicles, like many other designers, architects and engineers, lack general training in and knowledge of the fire performance properties of polymeric materials and the fire safety problems associated with them. *Recommendation:* Require increased emphasis on fire safety of polymeric materials in the scholastic curriculum for designers and engineers and maintain this emphasis through on-the-job training and review. Alternatively the design team should include one person who possesses the necessary knowledge of fire performance properties of polymeric materials.

Conclusion: Elastomeric polyurethane cushioning is widely used in transport vehicles; it burns readily even when fire retardants are incorporated in the resin. New materials and/or new approaches using current materials are urgently needed. *Recommendation:* Implement programs (chemical, materials, or engineering) to improve the fire safety characteristics of cushioning systems including consideration of smoke, toxic gases, flammability, and manufacturing and operational practicality. *Conclusion:* Carpet systems vary greatly in their response to a fire, and systems that are safe for uses in urban transit cars, buses, etc., need to be developed or defined. The use of carpeting on walls or ceilings of subway cars, buses, etc., is particularly undesirable. *Recommendation:* Develop an assessment technique to quantify the potential threat to the environment of additives, monomers, polymers, degradation products, and other materials used in the synthesis or modification of polymers to improve their fire safety characteristics.

Conclusion: Concern exists about potential fire hazards associated with the rapidly increasing use of polymeric structural and insulating foams in such vehicles as urban transit cars. *Recommendation:* Develop fire safety system analytical methods, including the use of scenarios, to guide materials selection and to permit overall fire safety assessment to be based on important factors such as material design, environment, detection, and fire control. *Recommendation:* Increase the development effort on char forming systems with particular emphasis on lowering the fabrication cost.

CHAPTER 5

TEST METHODS, SPECIFICATIONS, AND STANDARDS

5.1 Introduction

The roles of materials in fire safety are governed by a hierarchy of test methods, standards, specifications, codes, and regulations. In addition to the complexity of this system, the number of agencies involved presents serious difficulties both for the producers of polymeric materials and products as well as for the users. Ground transportation presents unusual problems; yet, although the number and types of vehicles are large and the total number of vehicles is enormous, only one federal regulation is officially in force concerning the flammability of materials.

While the objective of all fire safety activities is to reduce loss from unwanted fires, the precise definition of loss is itself both difficult and controversial. For the present purpose, attention will be focused on two major components — human death and injury and the destruction of physical property.

In ground transportation, test methods, standards, and specifications have their origin in the following activities:

1. Statutory efforts of government regulatory agencies to reduce the frequency and severity of fires through mandatory standards.
2. Voluntary efforts of associations concerned with the development of fire test methods, recommended safety practices, and the dissemination of authoritative fire safety information.

A test method is a procedure for measuring a property or behavior of a material, a product, or assembly as an aid to predicting performance. Specifications find their principal application in the definition of the properties of a material, product, or structure being procured. They establish the level of performance that the items must meet and the test methods by which such performance is to be measured. When incorporated into a purchase contract, the specification becomes part of a legally enforceable document. Ideally, a fire test method should be an experimental procedure that can be used to predict the performance of a material, product, structure, or system under fire exposure conditions which can reasonably be anticipated in the intended application.

Test methods may measure different *fire hazard* characteristics such as:

1. Ease of ignition.
2. Surface flame spread.
3. Heat release.
4. Smoke evolution.
5. Toxic gas formation.
6. Fire endurance.

Acceptance tests generally are classified according to intended end use of the product, which frequently is required to comply with specific regulations. Finally, tests frequently are classified according to size and designated by such descriptive and non-quantitative terms as large scale, full scale, subscale, small scale, and laboratory scale. From a fundamental standpoint, this is the least useful type of classification, but from a practical standpoint, the size of the test may be critically important.

A principal measure of the fire hazard in a particular transportation system is a series of tests that have been selected to evaluate the choices of materials to be used in that system. With results of such tests, design engineers may evaluate their choices of materials against the specification limits. The specification limits may be chosen to limit the choice to the few best materials or only to eliminate the worst materials in a given use category.

One of the most common criticisms aimed at a laboratory test method is that it does not fully simulate an actual, real-life situation. If a material were expected to be exposed to fire risk in one and only one precisely defined set of circumstances (size, orientation, type of ignition source, method of applying the ignition source, ventilation conditions, environmental conditions, etc.), then it would be obvious how to test for fire hazard. A series of candidate materials could be evaluated under this set of circumstances. One then could evaluate smaller samples of the same series of materials by means of another proposed smaller-scale test method, and the smaller-scale test method would be validated for future use if the results of the two procedures correlated.

Unfortunately, this idealized approach is frequently not directly applicable. The principal reason is that a material of interest may be exposed to fire risk in a wide variety of ways rather than by a simple, well-specified set of circumstances. Thus, an unacceptably large number of experimental fires would be needed to explore fully all of the permutations and combinations of the variables. A second reason is that, in some cases, even a few realistic fires would not be practical, if, for instance, each experiment required destruction of an expensive rail passenger car or bus or permissible if the experiment required exposure of humans to lethal fumes.

This chapter is aimed at enumerating and evaluating the current fire hazard test methods, specifications, and standards used in the selection of materials for ground vehicles. An additional objective is the correlation of these test methods with facets of fire safety covered in other chapters of this volume. In the several sections of this chapter will be found a listing of the regulations and specifications that have been adopted for use in the many different modes of ground transportation. (For a very thorough discussion of test methods, specifications and standards, the reader is referred to Volume 2.)

5.2 Organizations Involved in Regulations and Requirements

The U.S. Department of Transportation (DoT) was organized in 1967 from a number of separate government agencies, each of which had been in operation for some time. They are now called "Modal Administrations of DoT." Some of these agencies have regulatory power, but others do not. In this section, their jurisdictional activities, regulations, and requirements concerning flammability of materials in land transport vehicles are presented.

5.2.1 National Highway Traffic Safety Administration (NHTSA)

The NHTSA regulates the safety aspects of passenger cars, multipurpose passenger vehicles, trucks, and buses. It issues its regulations in the form of Motor Vehicle Safety Standards (MVSS). MVSS 302 applies to the flammability of interior materials and is a standard to which highway vehicle manufacturers must conform. (See Section 5.3.1.1 for discussion.)

5.2.2 Federal Highway Administration (FHWA)

The FHWA regulates the safety aspects of any road vehicle that operates interstate under its Bureau of Motor Carrier Safety. It has no regulations concerning flammability of polymeric materials but does regulate the number and type of fire extinguishers that must be carried by interstate buses and trucks.

5.2.3 Federal Railroad Administration (FRA)

FRA regulates the nation's railroads and controls the specifications for AMTRAK passenger vehicles. Specification requirements for flammability and smoke emission of polymeric materials are identical to those for the Urban Mass Transportation Administration (*vide infra*). Further, FRA has issued a "Guide for Preparing Accident/Incident Report." This guide requires that any transportation system operating railed vehicles shall report on all accidents or incidents that occur. By gathering such data over a period of time, FRA will be able to determine why changes are needed in regulations or what new regulations may be required.

5.2.4 Urban Mass Transportation Administration (UMTA)

UMTA is not a regulatory agency. It is provided with large sums of money by Congress to assist the cities to develop new or revive existing bus, rapid transit, and commuter rail systems. However, in providing these funds, Congress requires that system safety be a prime consideration. Since fire is a principal safety hazard in such systems, UMTA provides guideline specifications for flammability and smoke emission or materials for use in preparing design specifications for vehicles.

5.2.5 States and Communities

State and community governments do not regulate ground transportation vehicles insofar as flammability of the materials of construction is concerned. The transit authorities have mostly operated autonomously and set their own specifications for materials when purchasing vehicles. Where they now request federal government assistance in their purchases, they may use UMTA guideline specifications.

5.3 Specifications and Test Methods

5.3.1 Interior Materials and Furnishings

5.3.1.1 MVSS 302

As stated earlier, the only federal regulation for flammability of ground transportation vehicles is the Federal Motor Vehicle Safety Standard MVSS 302. This standard applies to seat cushions, seat backs, seat belts, head linings, convertible tops, arm rests, all trim panels, compartment shelves, head restraints, floor coverings, sun visors, curtains, shades, wheel housing covers, engine compartment covers, mattress covers, and padding for crash energy absorption.

The related test procedure is rather simple, requiring the use of a small metal cabinet. The test specimen is mounted horizontally in a frame, a 1½ inch Bunsen burner flame is applied at one end for 15 seconds and then removed. It is required that the material shall not burn or transmit a flame front across its surface at a rate of more than 4 inches per minute. However, if a material stops burning before it has burned for 60 seconds from the start of timing, and has not burned more than 2 inches from the point where the flame was when timing was started, it is considered to have *passed.*

Since this is the only federal regulation available for ground vehicles, many of the transit systems have begun to specify MVSS 302 in their procurements for new buses and rail rapid transit vehicles. MVSS 302 test procedure for combustible interior materials was deemed unsatisfactory by UMTA in providing for fire-safe mass transportation vehicles. UMTA, therefore, issued a document entitled, "Guideline Specifications for Flammability and Smoke" (see Appendix A for text).

5.3.1.2 Proposed Guideline Specifications for Flammability and Smoke Emission

These proposed guideline specifications apply to seat cushions, seat frames and shrouds, cushion fabrics, wall and ceiling panels, elastomers, plastic glazing, and lighting diffusers as well as carpeting and ducting as interior materials. The specifications are combinations of American Society for Testing and Materials (ASTM), FAA, and other test procedures, which, in some cases have been modified.

The test procedures and passing limits are as follows:

1. For seat cushions, wall and ceiling panels, seat frames and shrouds, plastic glazing, lighting diffusers and ducting, materials are required to be tested according to ASTM E162-67, Radiant Panel Test. In this test, the material is set on a rack at a 30° angle to a porous ceramic panel that has been heated to a surface temperature of 1238° F (670°C). The flame spread index (I_s) is a product of the rate of flame spread down the exposed surface and a rise in temperature of a thermocouple set in a stack above the apparatus. The standard is modified as follows:
   a. a 6-inch pilot flame is required.
   b. the sides and back of the sample must be covered with aluminum foil.

c. a screen must be placed under the sample.

d. flaming drippings are not allowed.

The I_s for plastic glazing and lighting diffusers should not exceed 100. Seat cushions should not exceed an I_s of 25 while the I_s for all other materials should not exceed 35.

2. For upholstery materials, the FAA/FAR 25.853 Verticle Method is employed. In this test, the material is suspended vertically, and a small Bunsen burner flame is applied at the lower edge for 12 seconds. The following modifications are made to this test:
   a. the average flame time after removal of the flame source should not exceed 10 seconds.
   b. burn length should not exceed 6 inches.
   c. flaming dripping should not be allowed.
   d. durability of any flame retardants is determined by subjecting to leaching by Federal Test Methods 191b Method 5830 or after dry cleaning according to American Association of Textile Chemists and Colorists-86-1968.
3. For elastomers, the ASTM C542 Standard is used. This test requires that the material be placed in a vertical position and a Bunsen burner flame be applied at the lower end. To pass this test, the flame should not propagate to the top of the sample after removal of the flame although the burner continues to be applied at the lower edge for 15 minutes.
4. For carpets (which must be tested with their underlay if one is used), the NBS Flooring Radiant Panel Test (NBS 1974) is invoked. The procedure and equipment are almost the same as in the ASTM E162 test, but the sample is laid horizontally and the radiant panel set at a 30° angle to it. The test continues until the specimen flaming goes out (extinguishment). The distance burned to extinguishment is converted to watts/cm². A limit of no less than 0.6 watts/cm² should be required by the guideline specifications.

5.3.2 Interior Materials (Other)

The remaining combustible materials for which tests are required are:

1. Thermal and acoustical insulation — These materials are also tested by ASTM E162 and should have an I_s no greater than 25.
2. Flooring — In most cases, the floor of a mass transportation vehicle is required to be a staunch fire barrier against fires that originate in electrical equipment under the floor. Thus, as required for fire barriers in building construction, ASTM E119-73 is used as a standard to test a floor's burn-through resistance. The sample is exposed to heat on a time-temperature curve. The floor should resist this heat for 15 minutes, up to a temperature of 1400°F (760°C).

   Furthermore, the heat should be applied to the underside of the structure, since many of the fires that occur in this type of vehicle come from below the floor.

5.3.3 Electrical Insulation

Ground transportation systems use two general types of electrical wire: light wire for lighting, control, public address system, auxiliary circuits, etc., and high voltage cable for power devices such as traction motors and regenerative braking systems. Two separate test procedures are called for:

1. Light wire is tested in accordance with Insulated Power Cable Engineers Association (IPCEA) standard S-19-81, Paragraph 6.19.6, with the added provision of FR-1. This test requires that the wire be suspended vertically in a small chamber and a small Bunsen flame applied at the lower end at a 20° angle for 15 seconds on and 15 seconds off for five cycles. The FR-1 modification provides that the flame not be reapplied until any flaming of test material caused by previous application of flame ceases of its own accord. Failure is signified by the charring of a piece of paper attached to the wire 10 inches above point of application of the burner.

   It is a serious lack that neither ASTM nor the Institute of Electrical and Electronic Engineers (IEEE) has a test to determine the integrity of this size of wire during flame impingement. It is a critical safety requirement that the functions served by these wires continue.
2. High voltage cable is tested in accordance with the IEEE Standard 383-1974. In this test, cables of 250 MCM size are suspended in a 96-inch (2438-mm) vertical rack and exposed to a 70,000 Btu per hour ribbon burner at a position 78 inches below the top of the rack. The cables are so wired that burn-through time is automatically registered for each cable. After 20 minutes of exposure or a total burn-through, the flame is shut off and the time that the flame continues and the length of damage to the cable are recorded. The guideline requires that the cables should retain their current-carrying integrity for 5 minutes after start of the test. (It appears likely that the heat flux will be increased to 210,000 Btu per hour, or even 400,000 Btu per hour, at some time in the near future to stimulate hotter fires that could impinge on electrical cable.)

5.3.4 Smoke Emission

The guidelines require testing of all materials listed above with the exception of foam seat cushioning, electrical insulation, carpeting, and flooring, using the NBS smoke chamber as well as the National Fire Protection Agency (NFPA) Standard 258, "Smoke Generated by Solid Materials." The allowable smoke density, D_s, in both flaming and nonflaming modes has the following limits:

a. For upholstering, air-ducting, thermal insulation and insulation covering, the D_s should not exceed 100 within 4 minutes after start of the test.
b. For all other materials, the D_s should not exceed 200 within 4 minutes after the start of the test.

An explanation of the exceptions noted above is in order. At the present time, foam seat cushioning that can pass the flammability test will not pass the smoke tests. Manufacturers are optimistic that this failing will be overcome soon; laboratory samples are encouraging.

A smoke test for electrical insulation awaits the formulation and acceptance of a method for performing this test. One group of experimenters propose wrapping 20-gauge wire around a special mandrel and testing it in the usual manner in the NBS Smoke Chamber. Others suggest use of a current overload as a means of performing this test in a manner consistent with the most probable heating of electrical insulation. Neither has addressed the problem of testing the insulation of heavy cable, especially where there is a composite construction. Work on the establishment of a satisfactory test procedure is now under way. In the interim, heavy smoking insulation should be avoided.

Carpeting is another category of materials that does not pass the smoke tests. An exception, however, is a recent innovation in carpeting using all-glass fibers, but the feasibility of the use of this type of carpeting in heavy traffic areas has not been established.

Flooring (structure) made of polymers using a honeycomb sandwich type structure, but without metal covers, appears particularly vulnerable to fire and smoke generation. On the other hand, there does not appear to be any reason to run a smoke test on flooring consisting of plywood faced with metal sheets on both sides since experience with its performance in fires has been good.

Before leaving the subject of smoke emission testing, recent work by Breden and Meisters (1976) at the National Bureau of Standards must be mentioned. They found that if thermoplastic materials are laid horizontally in the NBS Smoke Chamber and heated from above, much higher and more realistic values of D_s are obtained. It represents the results of a worst-case (and real case) condition for the material test. Materials that do not meet and flow are not much affected by this change in specimen orientation.

The entire list of UMTA proposed guideline specifications has been adopted for applications to AMTRAK passenger vehicles.

5.4 Discussion of Critical Elements in Flammability Testing

5.4.1 Test Geometry

For materials that are used in a variety of different orientations, a simple horizontal test is wholly inadequate. It is well known that the flame propagation rate in a vertically oriented sample may be more than ten times that of the same material ignited in the horizontal position. The MVSS 302 test therefore lacks significance. In the ASTM E162 test, the specimen is oriented at 30° to the vertical. This is an improvement but is not a worst case test.

The NBS Flooring Radiant Panel Test places the specimen in a horizontal position so that it is heated from above. It has been common practice to use the ASTM E84 Tunnel Test, which places the specimen on the roof of the tunnel while heating is provided from below.

5.4.2 Allowable Flaming After Exposure

Flame time after removal of the ignition source should be limited (e.g., to no more than 10 seconds) so as to provide a minimum opportunity for ignition of adjacent materials. Flaming drippings clearly represent opportunities for flame spread and should not be permitted.

5.4.3 Test Conditions

Flame spread rates and ignition times usually are determined on single, homogeneous specimens in the absence of any other heat flux to the specimen. Consequently, they may not represent the response of the material in a natural fire environment. In addition to tests for individual materials, components such as entire seats, possibly together with some adjacent components, should be tested.

5.4.4 Oxygen Index

The oxygen index test is a useful laboratory test to determine comparative values between materials of a particular category, but should not be used to determine material behavior in a real fire situation.

5.4.5 Heat Release

Rate of heat release may be defined as the heat produced by the combustion of a given weight or volume of material in a given period of time. This characteristic is relevant to fires in that a material which burns with the evolution of little heat per unit quantity burned will contribute appreciably less energy to a fire than a material which generates large amounts of heat per unit quantity burned in a given period of time. Over the past several years there has been growing support among those working in the fire field that this is an important criterion by which to evaluate the fire hazard from a particular material. The rate of heat release is a measure distinct from ignitability and surface flame-spread potential. Total heat release is another qualifying parameter.

DOT regulations at the present time do not provide for any standards or specifications on heat release. Such test methods need to be developed. (A description of heat release rate calorimeters is presented in Volume 2, Chapter 5).

5.4.6 Smoke Evolution

Smoke density may be defined as the degree of light or sight obscuration produced by smoke from a burning or pyrolyzing material in a given condition of exposure. This characteristic is relevant to fire safety because an occupant has a better chance of escaping from a vehicle if he can see the exit, and he is not poisoned or otherwise incapacitated by smoke ingredients. Measures of smoke density include the degree of light absorption and the specific optical density.

The NBS Smoke Density Chamber is the best equipment currently available for smoke evaluation; it has been adopted by the National Fire Protection Association (NFPA) as a national standard (NFPA 258). The ASTM also is expected to adopt it. Current practice shows a tendency for standardization on a single heat flux level (2.5W/cm^2); however, several laboratories are now conducting work to determine the effects of higher heat flux levels on smoke emission. The effect of sample orientation has already been noted (Breden and Meisters 1976).

Other test methods such as ASTM E84 and ASTM E162 provide a means for measuring smoke, but these are being supplanted by the method discussed above.

5.4.7 Toxicity

Chemical analysis per se provides only a limited description of the toxicity hazard of products of combustion. By this method alone, many organic toxicants would escape detection. A definitive test for toxicity must include an animal test that determines incapacitation of the animal and correlation to similar effects in humans.

work now being performed at the FAA Civil Aeromedical Institute shows considerable promise in the pursuit of a standard toxicity test. Reproducibilities of time to incapacitation and time to death on exposure of rats to gaseous products of combustion are excellent. Correlation also is good with the chemical analysis of gases produced.

Toxicity guidelines must be developed before interior materials toxicity requirements can be specified. An extensive discussion is presented in Volume 3.

5.4.8 Fire Endurance

Fire endurance may be defined as the resistance offered by a material to the thermal effects of fire. This characteristic is relevant to fire safety in that a material (and its related installation, detail, and structure) that has high fire endurance will contain a fire and provide more protection than one that fails to contain the same fire, all other factors being the same. Two measures of fire endurance are penetration time and resistance time.

The use of the ASTM E119 standard test for the determination of the fire endurance of floor structures of transit vehicles is a proper utilization of this method. Floors of such vehicles must act as the principal fire barriers to contain under-car fires.

5.4.9 Combustible Gas Evolution

Combustible gas evolving from burning of pyrolyzing materials may accumulate and produce flashover under given conditions of exposure. This phenomenon has been observed, for example, where polyurethane foam or foamed latex were used. The low temperature of thermal decomposition permits the production of large quantities of combustible gases by radiant heating of these foams. There are no adequate tests to completely define such combustible gas evolution.

5.4.10 Ease of Suppression

Ease of suppression may be defined as the relative ability of a particular extinguishing agent to extinguish a burning material. This characteristic is relevant to fire safety in that a material which is easily extinguished by a hand extinguisher or automatic extinguishing system presents less hazard than one which resists extinguishment.

With the current interest in completely automated mass transit vehicles that may have no on-board attendants, hand-held fire extinguishers may not be adequate for the protection of passengers. Automatic extinguishing systems such as Halon 1301 are useful, but the fire must be a "surface" variety subject to almost instantaneous extinguishment; otherwise, the decomposition products of the extinguishant become additional hazards. Experimental work in this area has been sponsored by UMTA. Tolerable levels of the extinguishing agent and its feasibility in real fire situations are being investigated. There are no current standard test methods for ease of extinguishment.

5.4.11 Prediction of Actual Fire Behavior

Prediction of actual fire behavior is made difficult by the large number of conceivable scenarios for any of the vehicles used in ground transportation. However, based on a knowledge of the behavior of materials in relevant laboratory tests and the design of vehicles, an experienced investigator can make a fair prediction of the outcome of a fire. Success of such a prediction is improved by further data obtained from large-scale compartmentalized burn tests.

5.4.12 Testing Adequacy for Systems

The proposed guideline specifications for flammability and smoke emission are a great improvement over MVSS 302. A recent trend in total system evaluation is to attempt definition of a total hazard index that might enable one to formulate an equation or mathematical model for a system which combines flammability, smoke, and toxicity hazards. This concept may be difficult to implement in part because a fully adequate definition of the hazard does not yet exist. The number and scope of variables in any given system presents a formidable challenge. However, this complexity does not preclude the use of a numerically defined gating profile for description of fire-related properties of materials (quantification of the analytical reflections of skilled observers in this field).

5.4.13 Modeling and Scaling Tests

Techniques for modeling and scaling for fire hazard are not sufficient at the present time to justify reliance on such techniques for ground transportation vehicles. Efforts to mathematically model related systems have not been successful because the number of variables is very large and the basic predictability of the rates of heat and smoke release and flame spread is distributed by many uncontrollable factors. Scaling, too, has been found to be largely unproductive.

5.4.14 Large-Scale Testing

Large-scale testing, as distinguished from full-scale testing, is being carried out at the present time by segmenting a full-size vehicle to permits its sections to be tested under conditions simulating a real fire. In performing these tests, one can determine the interaction of several different types of materials that normally are positioned adjacent to each other in the structure of a vehicle. It is essential that the real environment affecting the components be duplicated, including radiation effects and air convection.

5.5 Programs Needed for Improved Standards

Recent work has indicated that large-scale simulation of vehicle interior fires is needed for more accurate evaluation of the criteria for judging relative fire safety of various materials and systems. Research work should be performed to correlate mass availibility, area exposure, materials configuration, and ignition severity factors and parameters such as toxic gas evolution, smoke obscuration, and flame spread with the behavior of materials in vehicle interiors under real fire conditions.

To provide the data base for developing improved fire safety standards for vehicles, the following types of tests must be performed:

1. Basic property tests including, heat of combustion, rate of heat release, smoke release, etc.
2. Practical tests such as those listed under the guideline specifications.
3. Large-scale mock-up tests, such as compartmentalized tests that might include a set of seats and adjacent compartments or a similar set of components.
4. Full-scale fire tests of a vehicle to determine the effect of various design considerations on propagation of a fire.

To achieve earliest results, these tests should provide to the greatest extent possible for measurement of toxic gas evolution by employing direct animal response and analytical methods. The various parameters recorded through the smaller-scale tests should be compared and correlated with the relevant parameters in the larger-scale tests to determine the relative ability of the former to predict behavior under actual fire conditions.

5.6 The Adequacy of Materials Testing

Flammability test methods should have some common denominator useful for describing the probable combustible materials in a wide area of applications. However, misunderstandings and improper evaluations can occur unless the limitations of the test methods are understood.

To those unfamiliar with the details of fire tests, it would appear that the material which exhibits superior performance over another in one fire test would exhibit superior performance in any other. Unfortunately, this is not the case. Relative performance can vary considerably depending upon test used.

No single fire test, and perhaps no combination of a limited number of fire tests, can predict the behavior of materials under all possible conditions of fire exposure. Economic factors alone usually prohibit the performance of all fire tests that are relevant to all possible fire scenarios, therefore, fire tests can and should be classified according to intent. On this basis, they can be divided into three groups:

1. Laboratory research and development tests designed to generate information on the basic properties of a material or combination of materials and the effects of different variables on those properties. Such basic properties are exemplified by heat of combustion.
2. Pragmatic tests designed to simulate anticipated application conditions and intended to serve as standards on which specifications may be based. Examples include the federal flammability standards for carpets, mattresses, upholstered furniture, and children's sleepwear.
3. Full-scale or large-scale tests designed to reproduce actual fire scenarios under controlled and measured conditions. Such tests may be the only realistic basis for judging the validity of pragmatic tests used as standards for specifications.

Because of economic considerations, relatively few full-scale fire tests have been performed to validate the pragmatic tests being used as standards for specification. Regulatory agencies to a great extent have relied on previous fire experience to select the pragmatic tests needed for their particular situations.

Existing test methods available to regulatory agencies such as DoT are inadequate to provide complete guidance for the selection of polymeric materials to be used in systems that might face a wide spectrum of fire scenarios; such test methods are, at best, adequate for screening of individual materials whose fire characteristics are unsatisfactory. UMTA and FRA generally have shown good judgment in utilizing existing test methods with modifications and have achieved a reasonable degree of effectiveness.

5.7 Conclusions and Recommendations

Conclusion: Test methods available to regulatory agencies are inadequate to provide guidance for the selection of polymeric materials to be used in vehicles. *Recommendation:* Implement a major research program quickly to improve fire safety of vehicles and transit systems by developing the data needed to improve test requirements, test specifications, test data, and test extrapolation methods.

Conclusion: Only one federal regulatory standard for ground transportation, MVSS 302, applies to automobiles, trucks, buses, and recreational vehicles. This standard prescribes a test method that tests materials only in a horizontal orientation and is considered by test experts to be almost totally ineffective in providing for fire safety in a real fire situation. *Recommendation:* Develop new standards that will better define the fire performance of combustible materials in vehicles (e.g., standards recognizing that materials oriented vertically may spread flame an order of magnitude faster than the same material tested horizontally).

Conclusion: The flammability and smoke emission of materials used in public transportation vehicles such as rapid transit and railroad passenger cars currently is covered only by recommended guideline specifications, not by federal regulatory standards. *Recommendation:* Develop and implement rapidly regulations concerning allowable parameters for flammability, smoke emission, and toxicity.

CHAPTER 6

SMOKE AND TOXICITY

6.1 Introduction

This chapter briefly treats some of the more important smoke and toxicity considerations relevant to the fire safety of land transport vehicles. It is not adequate for a complete understanding of the problems of smoke and toxicity, and the reader is referred to Volume 3.

The principal hazards to survival of persons subjected to a fire environment are:

1. Heat destruction of tissues
2. Toxicity from oxygen deficiency and exposure to carbon monoxide and other noxious gases, aerosols, and particulate material
3. Presence of smoke with consequent reduction of vision and visibility
4. Fear or outright panic resulting in secondary mechanical trauma.

One or more of these four factors may be involved in a fire depending on the fire scenario and the individuals involved. The nature, shape, and amount of the materials undergoing combustion or pyrolysis determine the quantity of each of these factors involved in any fire and, therefore, the degree of hazard to human survival.

Modern subway cars are examples of a confined space in which fire represents a serious hazard. The material that undergoes combustion or pyrolysis consists of natural and synthetic polymers. In general, with the exception of the natural polymers (wool, cotton and other cellulosic materials), the preponderance of flammable materials are primarily synthetic polymers.

As indicated in Chapter 4, materials currently in use cover a broad spectrum. The choice of polymeric materials should permit modern transport systems to incorporate materials that meet or exceed current fire safety regulatory requirements. Flame-resistance requirements involve burn tests (e.g., the 60-second vertical, 10-second flame out, and "no flaming-dripping" tests). Regulating agencies now have few specific requirements concerning smoke production or noxious gas generation under fire conditions, but there is growing concern about these matters.

There is clear evidence that more deaths and injuries result from smoke and toxic gases than from the thermal effects of fire. Further, the long lasting effects of certain smoke and toxic gas injuries are not well known, particularly to the public.

Unfortunately, many toxic gases are neither odorous nor visible (in contrast to most smokes). Obscuration by smoke would not be expected to correlate with lethality; however, the particulate material in smoke may effectively adsorb highly toxic chemicals (such as HCl), thereby producing a toxic hazard. These facts are not adequately recognized by the public or sufficiently recognized in research or literature.

The predominant toxic product from fires is carbon monoxide; incapacitating or lethal amounts can develop within minutes of ignition.

6.2 Perspective on Experimental Data

The past decade has seen an increasing number of studies concerned with noxious gases and smoke resulting from thermal degradation of polymeric materials. Although the general implications of these studies is examined in Volume 3, particular points pertinent to vehicle fires are discussed below.

Pyrolysis or combustion products of the polymers used in passenger vehicle construction (railway and subway cars, buses, automobiles, etc.) have been found to include carbon monoxide (CO), carbon dioxide (CO_2), hydrogen cyanide (HCN), oxides of nitrogen (NO_x), ammonia (NH_3), hydrogen sulfide (H_2S), sulfur dioxide (SO_2), hydrogen chloride (HCl), hydrogen fluoride (HF) and phosgene ($COCl_2$) depending on the polymeric material. There is a substantial amount of data in the literature concerning the toxicity of these compounds. Other compounds are formed which are known to be highly lethal, but very little toxicological data about them exists or no current research is in progress (the physiological hazards of these gases are described in Volume 3).

The net physiological response from CO in combination with other thermal degradation gases is far from clear, although awareness of possible synergism is increasing. In real fire situations, heat, oxygen deficiency, smoke and panic may impose additional stress.

Fire situations can have an extremely complex toxicology. If a lethal CO concentration is not attained, other lethal or disabling factors may still be present; as for example in the series of experiments reported by Cornish and Abar (1969) about pulmonary injury from HCl developed in the absence of lethal effects from CO. A more subtle effect was noted by Effenberger in his series of experiments in which burning polystyrene did not cause rats to die or develop carboxyhemoglobin; rather depolymerization yielded styrene monomer that apparently had an immobilizing effect on the rats. This interpretation may be extended to fires involving humans, in which death became imminent because of gas-induced impairment of ability to escape from the fire. In the animal experiments cited above, the observation of incapacitation was based on performance in a swimming test, a simple exercise method also favored by Kimmerle (1974) to provide comparative data. Extension of tests involving animals is vital to meaningful progress in this area.

Smoke presents a number of hazards and a large variety of compositions. Smoke is basically a mixture of unburned carbon particles and other materials evolved by thermolysis and combustion. It may contain irritants adsorbed on the particles. The hazards of smoke may be both physical (blocking vision) and physiological (local or systematic chemical irritation and toxicity, heat injury, and panic). Volume 3 surveys hazards from smoke and describes current measurement techniques.

6.3 Experimental Data

In 1969, Gross, Loftus, Lee, and Gray reported their measurements of smoke produced during both flaming and smoldering exposures on 141 aircraft interior materials including sheets and laminate materials, foams, fabrics, rugs, pads, insulation, and films. The materials were similar to those used in land passenger transport vehicles; they were mostly synthetic and included polyvinyl chloride, acrylonitrile-butadiene-styrene (ABS), polymethyl methacrylate, wool, cotton, modacrylics, polyamide (nylon and aromatic types), polypropylene, urethane foam, polychloroprene (Neoprene) glass fiber, and paper. Smoke was measured by the progressive attenuation of a light beam passed through the smoke aerosol within an enclosed smoke chamber. Most of the materials produced more smoke during flaming, but certain materials produced significantly more smoke under smoldering conditions. All urethane foams produced large amounts of smoke under smoldering conditions. Neoprene, ABS, polymethyl methacrylate, and PVC materials nearly always produced more smoke under flaming conditions. This study illustrates the very wide range in smoke density values and specific toxic gases from materials used in a common location such as vehicle interiors.

A joint project by the FAA and NBS showed that a flash fire cell using a high voltage arc produced a flashover fire within 2 minutes with latex foam and polyurethane foam, as comapred to a flashover time of 3 to 4 minutes for polyethylene and acrylic resin and no flashover for PVC and cellulose.

An NBS project to study the burning characteristics of a Washington Metro bus and Metrorail interior was reported by Braun in December 1975 and by Birky in February 1976. The projects involved ignition of interior furnishing materials and observation of the spread of fire and smoke. Analyses also were made of the gaseous products of decomposition. During the Metrorail car test, the effect of the gases on animals was studied. Full-scale tests were run; a mock-up of a section of the car's interior was constructed for the animal test. Some of the more significant results were:

1. In comparing smoke data from three full-scale tests, the most critical event was the ignition of the foam (polyurethane) padding in the seat assembly since zero visibility occurred about 1 minute later.
2. A toxicological hazard definitely resulted from the combustion of polyurethane foam cushions. The hazard arose from the foam cushioning rather than some other portion of the seat. The test was inconclusive as to whether the toxicological hazard was created by the observed increases in carbon monoxide or hydrogen cyanide.

Although very few large-scale tests have been performed on mass transportation vehicles, there have been a number of fires on various types of vehicles in different circumstances that have provided valuable insights. For example, in an actual vehicular tunnel fire (Boston, 1975) the combustion of only a comparatively small amount of polymeric material provided a large amount of smoke. Since the vehicle effectively blocked tunnel ventilation, these smoke concentrations produced zero visibility.

Modern tunnel construction attempts to deal with this problem by the judicious placement of fans in exhaust ducts. These fans often are directed from a central control point so that they can be activated according to the location of the burning vehicle.

6.4 Clinical Data Based on Vehicle Fires

Quantitative toxicological data based on land transport vehicle fire victims is almost nonexistent. Even data about aircraft fire victims are limited. Smith and associates have described the results of forensic investigations of aircraft casualties as follows:

Two commercial aircraft accidents in the United States (Denver, Colorado, 1961; Salt Lake City, Utah, 1965) contributed greatly to the initiation of the present concern over the toxic hazard of the gases generated and fires. These accidents were of special significance because careful analysis indicated that few, if any, of the occupants would have significant physical injury from the relatively mild impacts involved; yet, a total of 60 persons perished as a result of thermal and chemical injuries sustained in the ensuing fires.

Carboxyhemoglobin measurements on 16 victims of the Denver crash revealed CO concentration ranging from 30 to 85 percent with a mean of 63.3 percent. Similar analyses on 36 victims of the Salt Lake City accident yielded CO concentration ranging from 13 to 82 percent, the mean being 36.9 percent. The lower carboxyhemoglobin values found in the second accident among the victims who were unable to egress indicate that the survival time for many victims must have been shortened by direct thermal effects or by toxic gases in the cabon other than carbon monoxide. There is a growing body of evidence to support this assumption.

In 1970, blood samples from victims of an aircraft crash followed by fire (Anchorage, Alaska, 1970) were analyzed for the presence of cyanide; the first time, to the best of this committee's knowledge, that such an analysis had been made on victims of an aircraft fire. Measurable amounts of cyanide were found in 18 of the 19 specimens tested. Concomitant carbon monoxide concentration ranged from 17 to 70 percent. In the one sample in which cyanide could not be detected, the carboxyhemoglobin concentration of 4.9 percent did not exceed a level that could result from smoking, indicating the probability of death on impact. Blood cyanide levels in these victims correspond closely with those reported in the literature for victims of structural and vehicular fires ranging from the lower detection limit (*circa* 0.01 μ g/ml) up to 2.26 μ g/ml. The relationship between cyanide levels and carboxyhemoglobin content varied in random fashion, perhaps representing relative proximity of the victims to cyanide producing materials. Alternatively, the varying cyanide levels reported may be due to uncontrolled autoproduction of cyanide in and from the tissues. Fires in new railway cars, subway and other rail vehicles and buses may be expected to show similar results if passenger egress were blocked.

6.5 Clinical Aspects of Toxic Fumes and Smoke

These matters are highly complex and require detailed examination and discussion. They are discussed in Volume 3.

6.6 Evaluating the Hazard of Toxic Fumes and Smoke

6.6.1 Comparison of Material

Valid comparisons of different materials must be based on similar or reasonably standard conditions. The desirability of a standardized test procedure may be questionable, but the principles of toxicological testing must be observed if any reasonable comparison of toxicological hazard is to be made. Real fires possess two essentially uncontrollable variables — oxygen supply and temperature — that makes selection of such standardized test procedures inherently difficult. Generally, laboratory thermal degradation tests in an oxygen-lean atmosphere are described as pyrolysis tests. Combustion tests, on the other hand, are conducted in oxygen-rich atmospheres. Either pyrolysis or combustion can be the more hazardous depending on the nature of the material being tested. Procedures must take both categories into account, either separately or together.

6.6.2 Thermal Decomposition Temperatures

Species and quantities of combustion products in a real fire situation can vary dramatically with the temperatures to which materials are subjected. This is an area in which more research is urgently needed. The continuing development of new synthetic polymers increases the complexity of the situation.

6.6.3 Analysis

Testing to identify the chemical components involved in gas emissions can help in understanding the effects of altering variables such as temperature and oxygen. The relative hazard of lethality of the combustion products can be estimated with reasonable confidence if a single component, such as CO or HCl, is clearly predominant and no other significant source of stress is present. However, if a significant quantity of other gases, heat, or smoke are generated, the net physiological response is difficult to estimate. Analyses of such mixtures or their degradation products are becoming less difficult with the development of new, more sophisticated analytical tools. Unfortunately, the resources and skills applied in this area of research are insufficient to make desired progress, particularly in the light of the fact that fabricated products may contain varying amounts of polymer, anti-oxidants, fillers, additives, and finishes that may be combined in many ways.

6.6.4 Testing

Substantial additional effort is required to develop suitable methods and procedures for determining the nature and characteristics, particularly lethal and incapacitating effects, of the fire products of polymers currently in use. A safer product from the standpoing of a flammability test does not always result in a truly safer product from a toxicological point of view because:

1. Polymer molecular structural modification or use of additives may result in the intended retarded ignitability of the product; but, when exposed to external heat, increased toxicity of combustion product or dense smoke evolution from the smoldering of the product may result.
2. Many of the more common fire retardants contain halogens, such as chlorine and bromine, that could contribute to the production of thermal decomposition products, e.g., hydrogen chloride (HCl), phosgene ($COCl_2$), and hydrogen bromide (HBr) in a developed fire.
3. The presence of nitrogen atoms, either in a polymer or in the additives, introduces the possibility of thermal degradation to HCN or NO_x.
4. Phosphorus-containing fire retardants may yield physiologically potent phosphoric esters when thermally degraded.
5. Other toxic species may be evolved.

Thus, any assessment of total fire safety must include some measure of toxicity and smoke in addition to flammability.

6.6.5 Extension of Analysis and Testing

Present procedures include combined testing, predictive testing, and "compartment" or large-scale fire tests.

Although correlation between gas analysis and effects on animals has been established, it would be hazardous to infer that one only needs to analyze for CO, HCN, HCl, etc., to understand the hazard. Animal tests must be carried out to detect the possible presence of a lethal species that has not been found analytically as well as synergistic effects. In fact, animal tests appear to be the only available definitive method for determining the toxicity of gaseous combustion products.

6.6.6 Epidemiological Studies

Analysis of epidemiologic data has been applied to fire toxicity only recently. There is a need for evaluated reports from passenger vehicle fires that provide comprehensive casualty data, including quantitative toxicological and pathological evaluation of victims. Such casualty data have started to become available. A recent discussion of HCN (as a suggested major lethal factor in aircraft fires) indicates not only current interest and concern in this area, but also the difficulties that may be associated with such analysis.

6.7 Conclusions and Recommendations

Conclusion: As the diversity and amount of polymeric materials used in vehicles increase, the problems presented by the generation of smoke and toxic gases in a fire also increase. Objective information defining the extent and nature of this hazard is not available. *Recommendation:* Increase in scope as well as effort and closely monitor for applications to land passenger vehicles the research program established to develop criteria and practices for determining the degree to which polymer flammability and decomposition products contribute to human morbidity and mortality.

Conclusion: Although it is known that thermal decomposition products from synthetic polymers contribute to the overall hazard during vehicle fires, the degree to which each product contributes is difficult to establish because the relationships of chemical and physical composition to smoke and toxic gas formation during combustion are not well understood. Species and quantities of pyrolysis and combustion products vary, in most cases, depending on the temperature to which the materials are subjected and the amount of oxygen available.

Conclusion: Carbon monoxide is a well established major toxic hazard in polymer fires, but current data suggest that, under both clinical and experimental conditions, thermal decomposition products other than CO can be major contributors to the hazard to human survival. Smoke and toxic gases are often more dangerous to human health and survival than the thermal effects of combustion (burns). Some toxic gases are not discernible by normal human senses. *Recommendation:* Develop and implement an educational program to advise the public and emergency services personnel of the increasing hazard that smoke and toxic gases pose to human health.

Recommendation: Expand research program directed at assessment of the hazards from toxic products to more clearly and rapidly define the effects of varying temperatures and oxygen availabilities during combustion of polymer materials.

CHAPTER 7

FIXED GUIDEWAY SYSTEMS

7.1 System Design, Operation and Fire Safety

7.1.1 Introduction

A fixed guideway system may be defined as a transportation system, either electrified or having on-board self-motive power, that utilizes guidance means involving positive mechanical contact with a fixed way, operates on a right-of-way for the mass movement of passengers either intra- or inter-urban, and consists of its fixed way, passenger vehicles and other rolling stock, power system, stations, maintenance facilities and other stationary and movable apparatus and equipment, and its operating practices and personnel. Under this definition are included rail rapid transit, rubber-tired rapid transit, commuter cars, street cars, and railroad cars.

Rail rapid transit is a high-frequency, high-capacity rail system operating on an exclusive, grade-separated right-of-way, whether at grade, in subway, or on an elevated structure. Many systems use all three modes, riding on the surface in the suburbs and using either tunnels or elevated structures for more traffic-congested areas. No *rubber-tired rapid transit* systems exist in the United States, but such systems are employed in Montreal, Paris, and Mexico City.

Rapid transit vehicles are electrically powered and may operate in trains of up to 12 cars. They take their power (600 to 1,000 VDC) from third rails or from overhead catenaries. High-level platforms and multiple doors on each vehicle provide for rapid loading and unloading at stations.

Commuter rail is defined as an urban rail passenger service, typically operated by inter-city railroads within 30 to 60 miles of central cities. Equipment may be diesel or electric locomotives, hauling passenger coaches, self-propelled rail diesel cars, or electric self-propelled multiple-unit vehicles. Where the equipment is electrically powered, current is collected from a wayside power rail or an overhead catenary.

The right-of-way is exclusive but not necessarily grade-separated. Commuter service usually shares the same facilities with intra-city freight and passenger service. Operations are governed by normal railroad procedures and work rules.

Light rail transit (LRT), colloquially sometimes called a streetcar system, is a rail guideway system wherein the route configuration may include non-grade-separated portions. It may operate in city streets with vehicular traffic or in reserved or median strips with vehicular crossings at intersections. The light rail vehicles (LRV) are electrically powered, are capable of operating singly or in trains, and can be constructed to accommodate loading from either high or low platforms or street level. The Massachusetts Bay Transportation Authority (Boston and vicinity), for example, operates streetcars that travel both at street level and in a series of tunnels with stations underground. LRT electrical power usually is collected from an overhead catenary by pantograph or trolley pull.

Railroad systems operate long distances between cities on exclusive grade-separated rights-of-way. Equipment may be diesel or electric locomotives and hauling passenger coaches. Where equipment is electrically powered, current is collected from an overhead catenary using a pantograph.

Although these systems differ from each other in their modes of operation, the polymers used in their construction and furnishing are similar. The following discussion indicates how their modes of operation create particular fire hazards which are unique, in what manner they share the same hazards, and in what ways they differ.

7.1.2 System Design Consideration from a Fire Safety Standpoint

7.1.2.1 Introduction

This section provides a guide for the design of polymeric components insofar as the material selection is relevant to fire safety. Considerations for material selection are discussed as they pertain to some hardware parameters controlled by the designer. These parameters include consideration of part function, geometry, and location and the influence of the environment in which the composite will function.

7.1.2.2 Component Considerations for Materials Selection

7.1.2.2.1 Part Function, Geometry, and Location

The accepted approach to satisfying a needed component function is through establishment of detailed design criteria and specifications. The designer then can initiate the design process and consider trade-offs to achieve the most efficient design and utilization of the most effective materials. If satisfaction of the part functional design requirement(s) cannot be achieved, any remaining desired criteria become academic. This concept establishes the sequence of design flow for any hardware item, subassembly, or assembly in the system regardless of other requirements or desirable features and provides a timely basis for further trade-offs in the selection of material. A typical sequence that might be established for candidate parts or subassemblies, covering design relative to function and fire characteristics is presented in Figure 1.

The geometry of the vehicle underbody and interior subassemblies — underbody electrical insulation and conduit, pneumatic and hydraulic hoses, air-conditioning ducts, as well as the interior seats, floors and carpeting — has a very important effect on fire safety. For example, the tendency for build-up of gaseous decomposition products and their subsequent explosive ignition (flashover) may be affected by the location of the materials exposed to ignition. Flame propagation (or spread) depends on the path of the flame front (i.e., horizontal, vertical, or at some intermediate angle). These geometric considerations are extremely difficult to define; yet, the designer must consider them carefully in part or subassembly design.

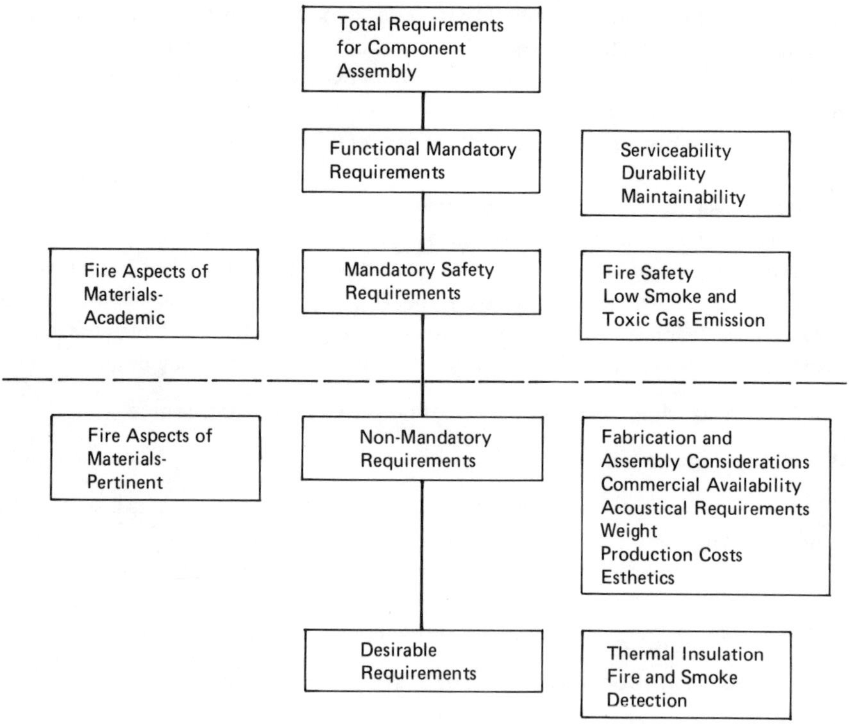

Figure 1. Flow diagram for prototype componentry.

The location of a specific part or subassembly also can be a factor in materials selection. Materials that melt and drip, for example, are hazardous as overhead interior components since they can ignite other compounds or fall on occupants.

Elevated temperatures experienced in fires, the rate and quantity of heat input, location in system, and thermal energy transition to the ambient are also important parameters. Flammability tests show that burning of a material after ignition is accelerated as the ambient temperature increases. A typical example is wool carpet. When wool is tested per FAA FAR 25.853a, in room temperature air, the flame time is 15 seconds and burn length is 3.6 inches. When the temperature of the air in the chamber is raised to 250°F (121°C), the flame time is 2.4 seconds and the sample is completely consumed.

7.1.2.3 Part Versus Assembly

The designer must consider the abilities of materials to withstand both ordinary and abusive wear without significant change in flammability characteristics. In a system intended for continuous use over long periods of time, the durability of the materials is of primary concern.

The evaluation of a material for fire behavior requires consideration of the material in the assembly or part in which it will be used. Adjacent assemblies also must be considered and the effects of material choices on the entire assembly or system must be established.

Parts or subassemblies cannot be averaged to obtain their combined fire characteristics but must be evaluated as complete assemblies. In most cases it is not economically feasible to determine the effect of a particular ignition source on an entire vehicle. Except as will be indicated below in Section 7.1.4.2, a vehicle may be segmented or "compartmentalized" into systems of materials and their fire behavior in full-size tests extrapolated to the total system. Tests of this kind have been performed by UMTA and will be undertaken by FRA for AMTRAK vehicles.

In the selection of materials that may pass a given set of specifications, there will be a wide range in test results. Materials then may be "ranked" in order of their effectiveness in flame retardance, smoke and toxic gas emission, etc. A risk assessment might be required to determine whether a particularly hazardous condition is created if the lowest ranked material that passed in each category is selected for inclusion in a single vehicle.

7.1.2.4 The Tunnel Fire

Several types of transit vehicles use tunnels for a significant portion of their total travel. Rail rapid transit vehicles use tunnels for a major portion of their trackage and are most familiar to the public in this form. Street cars like those in Boston also use tunnels and even have many stations underground. Railroad passenger trains use tunnels in traversing the countryside and frequently have tunnels up to a mile in length under hills. Trains sometimes have a station in a tunnel (e.g., Grand Central Terminal in New York City). As more air rights are sold over train tracks and roadways, the right-of-way increasingly resembles tunnels. Many of the fire hazards that are associated with conventional vehicular tunnels will be present as will added dangers to the structures that have been built over them.

The principal fire hazard consideration concerning tunnels is the difficulty of egress in an emergency situation and the difficulty of firefighting. A tunnel fire has all the elements of an oven — a confined space, radiative heating, and limited ability to dissipate smoke. Emergency accesses through which passengers can escape to the surface and emergency crews can reach the scene of an accident are located at intervals in tunnels under streets. A tunnel beneath a river or harbor, however, has only two exits, one at each end, and some of these tunnels are miles long. Tunnel fires are considered the "worst case condition" when predicting the behavior of materials in a fire situation.

7.1.3 Ventilating Systems

Design and operation of vehicle ventilation systems have important effects on the fire dynamics involved in a vehicle fire. Fire dynamics, in turn, determine the distribution of heat, smoke, and species of toxic gases in the vehicle. Of perhaps equal importance is the design and operation of the ventilation system of a fixed guideway system itself, particularly in connection with subways or operations in tunnels.

Few systems have been designed giving fire safety concepts adequate consideration; most have no closures or special operating procedures for a fire situation. In some cases, ventilation operations cannot be changed from within the car (in case of fire), and in such situations, the polymeric materials used in vehicle construction, and particularly their behavior in fire, become an important element in public safety and survival.

Most of the ventilating system types used in fixed guideway vehicles include some form of heating while some include air-conditioning as an integral function. Such systems usually involve:

1. Simple "once through" forced air with exhaust via available vehicle openings.
2. "Once through" forced air with a separate exhaust system.
3. Recirculation type forced air, with recirculation amounting to one third or to two thirds of the air removed.
4. Recirculation of air with air-conditioning added.

Air usually is sucked (or scooped) into the vehicle by a fan(s); passed through filters, if provided; mixed with recirculated air (in recirculating type systems); and then passed through a ducting system or through an overhead or wall plenum to distribution points in the car. Exhaust openings in the car collect air and pass it through ducting, and exhaust fans discharge it to the outside. In recirculating systems, only a portion of the air is discharged and the rest proceeds through an air-conditioner (or other treatment device) to a mixing chamber where fresh make-up air is added, and the combined recirculated-plus-make-up air is sent through a delivery ducting system into the passenger spaces. In older systems, ducting and plenums generally were of metallic construction, but newer systems increasingly feature polymeric materials in lieu of metals. These polymeric materials represent a wide range of compositions and combinations, and a wide variety of finishes is employed.

In a fire situation, "once through" systems could remove heat, smoke, and toxic gases, thereby increasing survival time for car occupants. Conversely, make-up recirculating systems can distribute heat, smoke, and gases to all portions of a car, *including those not involved in the fire.* Ventilating systems using polymers for ducting, etc., could not only burn and fail to perform designed functions, but also could contribute to the fire and the potential toxicity load. Thus, it is clear that the design, construction, and operation of car ventilating systems for fixed guideway vehicles should be a major determinant of the types and amounts of polymeric materials used in car construction.

Similarly, ventilation as a design safety feature in a subway or any other tunnel environment is an extremely important consideration. After a fire occurs, particularly in a car, there is no place for the heat, smoke, and toxic products of combustion to go, and an atmosphere unsuitable for human survival is created. Modern rapid rail transit systems have recognized this problem and built ventilation systems with fire safety features (the routine function of a ventilation system is to equalize temperatures in the tunnel and relieve air blast from trains entering stations). Ventilation devices vary greatly from open grates (or nothing) in older systems to sophisticated temperature/remote-control-powered ventilation equipment in newer systems. The design configuration of a modern system typically has a gravity vent shaft near the ends of stations with power fan shafts between stations. Longer sections of tunnels may have intermediate vent and fan shafts. Trains themselves provide the primary source of ventilation through piston action, but should the temperature in a tunnel reach an unacceptable level, the fans and vents work in concert to increase air movement. Normally, thermostats activate the fans and open vents when tunnel air exceeds a pre-set temperature.

In an emergency, the system control center has the option of controlling the fans remotely. It may run any fan in an exhaust or supply mode, but when the fans are actuated, the louvres in vent shafts automatically close to draw fresh air through the stations. Manual controls at the fans also can be used.

Control of the fans for emergency ventilation must be carefully planned in anticipation of an emergency. Several scenarios for each tunnel section must be developed and contingency plans written. In addition to transit system personnel, the local fire department should have direct input into the development of an emergency ventilation plan since the plan must not only protect car passengers and firefighters, but also the public where affected by system ventilation exhaust. The design of transit system ventilation equipment also affects the choice of polymeric materials used both in tunnels and in the cars. Ventilation system fire safety concepts (as they affect choice of polymeric materials) are extremely important in attended fixed guideway vehicles but vital in unattended fixed guideway vehicles (whether or not they operate in subways or tunnels).

In Section 7.1.4.1, an example of a fire scenario is provided. In Sections 7.1.4.2 through 7.1.4.6, the important elements of fire scenarios of fixed guideway vehicles are discussed.

7.1.4 Fire Scenarios As a Tool for Design Review

As noted in Chapter 3, the selection of materials for use in a transit system requires an intimate knowledge of transit fires and the fire behavior of materials. An understanding of fire dynamics, coupled with the use of appropriate fire scenarios, can help provide such knowledge.

In presenting a fire scenario (e.g., of a fire occurring in a vehicle in a tunnel), it must be understood that the scenario, while specific, leads to generalization. The particular situation can change radically depending on many factors (i.e., materials of construction and furnishing of the vehicle, specific application of the materials, placement of components, ignition source, passenger load, configuration of the tunnel, tunnel air convection, presence and use of tunnel exhaust systems, presence and use of extinguishment systems, and degree of training of the vehicle and emergency crews); however, valuable, generally applicable lessons can be obtained from each fire analysis.

Careful analysis of the fire accident reports (if the events have indeed been reported with sufficient detail and accuracy) can provide information to the designers of vehicles or entire systems that may make a difference in the degree of hazard built into a vehicle or system.

The importance of scenarios to developing an understanding of the several stages of a fire situation (from ignition to final suppression) cannot be overstated. Before completing a vehicle design or refurbishing a transit vehicle, competent and experienced engineers, operating personnel, materials experts, and other cognizant individuals should analyze the pertinent drawings or mock-ups with the question: What would happen if . . .? The scenarios thus generated would pinpoint most of the faults that could lead to fire hazards.

7.1.4.1 Example of a Fire Scenario — A Subway Fire

A three-car streetcar train was entering a subway station when the station starter noticed smoke coming from under the last car. He evacuated passengers and extinguished the traction motor fire with an extinguisher. The train was removed from the station, but the following train ran into a hanging live trolley wire that had not been noticed. The wire arced a hole in the front panel of the first car, then fell onto the roof where it arced a hole in the metal roofing and ignited combustible materials in the ceiling of the car (Figure 2). The first car continued to burn, and radiant heat soon ignited the roof of the second car of the train. Due to the resulting confusion, the fire department was not informed for about 40 minutes. When the fire apparatus arrived, the smoke was so dense that visibility was virtually zero. After the fire was extinguished, empty cars were run through the tunnel to drive the smoke to the exit a few hundred feet away.

Two hours later another live trolley wire was found hanging approximately 400 feet behind the original break. This wire rested against a wooden car door and therefore was not grounded (in all, 1,500 feet of trolley wire were replaced.).

Using this fire scenario it was learned that:

1. The trolley and the traction motor in the first car had so grounded the trolley wire that a very heavy current flowed through the wire, overheating and softening it until it broke.
2. The metal roof of the following car provided no insulation against a live trolley wire dropping on it.
3. The tunnel itself had no standpipes to supply water for extinguishment, and hoses had to be brought in over a long distance.
4. The tunnel had no fans for evacuating the smoke.
5. Although the passengers were evacuated safely, the smoke was intense and 39 firemen and transit personnel were treated at a hospital for smoke inhalation.

Scenario analysis stimulated a number of actions:

1. The tunnel was renovated to provide water standpipes and exhaust fans.
2. The car roofs were provided with a thin insulating layer sufficient to prevent future grounding of trolley wires.

Figure 2. Subway fire.

3. Transit personnel and the fire department were instructed to coordinate their activities so that a potentially dangerous delay would be avoided in the future.

7.1.4.2 Prefire Initiation Considerations

7.1.4.2.1 Determination of Fire Potential

An analysis of fire accidents in transit vehicles indicates a variety of ignition sources that, in consort with the fire loads present in transit vehicles, provide the potential for fire in various locations throughout the vehicles. Engines, motors, electrical wires and components, food service areas (on trains), cigarettes, and vandalism appear to have the greatest potential for ignition.

Fires that originate in cars mostly are the result of electrical wire malfunctions or vandalism. Car design and materials selection can minimize these sources of fire. There is no reasonable way to control the fire load originating in passenger carry-ons such as newspapers and parcels.

Ignition of polymeric materials is an extremely complex process. It depends on the nature and characteristics of the ignition source, on the availability of adequate oxygen, and on the physical and chemical properties of the polymer. Other important properties of the polymer influencing its ignitability include thermal conductivity density, specific heat, and activation energy. Once ignited, some polymers burn intensely, releasing large amounts of energy. Typical heats of combustion for various materials are listed in Table 1. This energy is available for transfer by radiation and convection to other combustible materials and propagates of a fire.

Table 1. Material Fire Load Data

Heats of Combustion for Typical Materials Found in Vehicle Passenger Cabins

Material	**Btu/lb**
Rayon	6,700
Tobacco	6,800
Cotton	7,100
Acetate	7,700
Polyvinyl chloride	7,700
Triacetate	7,800
Wood	8,800
Wool	9,000
Polyester	9,000
Modacrylic	11,000
Unsaturated polyester	13,000
Nylon 6	13,000
Spandex	14,000
Foam rubber	15,000
Bituminous coal (for comparison)	15,000
Urethane	16,000
Polystyrene	18,000
No. 6 fuel oil (for comparison)	18,000
Butadiene/styrene copolymer	20,000

Many polymeric materials burn with retention to structural integrity while others melt and sag. The latter materials may represent a greater hazard when they are used in a load-bearing application rather than a decorative one. Other properties such as melting and dripping, smoke evolution, rate of heat release, and burning rates are additional concerns that can be evaluated only in a finished product.

Fire potentials of individual materials have been ranked somewhat by small-scale static tests. The fire potentials for combinations of materials in use in a real environment have not been adequately determined and there is no acceptable predictive methodology (e.g., computer modeling, scale modeling, characteristics modeling).

7.1.4.2.2 Passenger Compartment Ignition Sources

An understanding of ignition sources and the probability of fire development is vital to proper evaluation of materials. Temperatures achieved by small heat sources (e.g., cigarettes, matches, lighters, etc.) are sufficient to ignite synthetic polymers. The following are some typical measurements of small ignition sources:

1. Cigarette, no draft, 1,050°F (565°C).
2. Cigarette, draft, 1,350°F (732°C).

3. Paper match, no draft, 1,508°F (820°C).
4. Cigarette lighter, no draft, 1,200 to 1,500°F (649 to 816°C).
5. Ignited newspaper, on or under a seat, 1,300°F (700°C).

Electric wire passing through or around the passenger compartment may cause ignition due to an overload current, fraying of the insulation, or breaking of the wire. Insulation of the wire should be selected with regard to flammability and the products of combustion. It should further be required that a non-conducting char or inorganic coating remain on the wire after direct flame impingement so that wire integrity will not be destroyed. It is essential in any fire situation that important functions such as lighting, control, and public address systems remain operational.

7.1.4.2.3 Undercar Ignition Sources

Statistics indicate that approximately three-quarters of all rail transit system fires originate under the car because of brake or hydraulic system failure, shorting of electrical components, or arcing to the power rail. Electric shortage batteries also have been a source of undercar fires.

Insulators for the power collector shoe are made of wood or reinforced plastic. If cleaning maintenance is not adequate, accumulations of oil, grease, and metal particles gather under the car and particularly on the power collector insulators. At some point, arcing of a collector on the power rail can cause ignition of accumulated grime. This form of ignition rapidly involves electrical insulation, which is notoriously smoky. As mentioned above, it takes very little smoldering of burning material to provide considerable smoke, and in a tunnel situation, even a small fire of this sort is a serious hazard.

7.1.4.2.4 Wayside Ignition Sources

Papers and other trash accumulate readily around rails. Such debris, as well as oil, grease, and metal particle accumulations, in combination with dry wooden ties is easily ignitable. In older transit systems, light wire and cable are exposed along the track and easily become involved in track fires ignited either by third-rail arcing, vandalism, or careless use of cigarettes or welding equipment. In new systems, the cables are carried in conduits within the tunnel walls. Many transit systems perform periodic cleaning or vacuuming of the tunnels. This has been very effective in reducing the number of fires from wayside ignition sources.

7.1.4.2.5 Wall Radiation Fire Enhancement

Radiation of heat from the walls and ceilings of tunnels has been largely ignored when calculating the heat modes effects of materials used in vehicle construction. Heat flux on these materials is greatly intensified by radiation so that they burn faster and other combustibles become more easily ignitable. Radiation also aids in spreading fire from a burning car to others in the train. This factor should be of prime consideration in specifying materials for use in vehicles that spend any significant portion of operational or storage time in tunnels or buildings.

7.1.4.3 Fire Load

One of the classic approaches to understanding the potential severity of fires in a given space has been to measure the fire load or potential heat release under fire conditions. Every potential fire in a vehicle (i.e., one that could occur at any time or place whether the vehicle is in operation or storage, and with full load or empty) must be considered as one in which all combustibles will be completely consumed. The question that must be answered is: What will this total heat load do if it is released in a relatively short time, at any time? Table 2 indicates the potential fire load in a rail transit car.

Table 2. Rail Transit Vehicle Combustible Materials (Typical).

Application	Mateial	Approximate Weight	
Wall and ceiling liner	PVC blends	2,291	lbs
Seat cushions	Polyurethane foam	640	
Carpet	Wool	425	
Carpet pad	Latex foam	265	
Partitions/wind screens	Melamine-faced wood	2,390	
Floor panels	Metal-faced wood	175	
Window panes	Acrylic	337	
Lighting fixture lens	Polycarbonate	45	
Air ducts	PVC blends	30	
Ceiling insulation	Polyurethane foam	307	
Cab shell	Epoxy/fiberglass	283	
Window glazing strips	Chloroprene & CRS Rubber	308	
Sealants	Polysulfide	115	
Electrical insulation	Chlorosulfonate polyethylene	436	
	Polytetrafluoroethylene	43	
Mis. interior trim	PVC blends	77	
	Epoxy/fiberglass	120	

7.1.4.4 Personnel Danger

In any vehicle fire, the operating personnel and firefighters as well as the passengers are in danger. Hazards faced by each of these groups are somewhat different. Passengers are unfamilar with transit operations and firefighting. Operating personnel usually are not trained firefighters. Experience has shown that firefighters often do not have sufficient knowledge of transit system operations and materials and, thus, are exposed to unrecognized dangers.

7.1.4.4.1 Passenger Egress

Fires are particularly traumatic experiences. Even when the means of leaving the danger site is relatively easy, panic frequently immobilizes some people and can spread very rapidly. Firefighters often find victims only a step or two from a door or window leading to safety.

Operators have described panics in vehicles traveling at ground level where egress is relatively simple. How much more trauma is there in a vehicle stopped in a tunnel, without lights and with smoke present?

Several conditions are of prime importance to passenger safety:

1. Vehicle materials of construction and furnishings must be chosen to maximize time for passenger evacuation.
2. The vehicle must be designed to provide as many means of egress as possible, consistent with structural integrity. Such exits would include doors that can be opened by passengers, windows that can be pushed out, or removable hatches in the roof (in case of vehicle overturn). Passenger openable doors or windows must be traded-off with the risks introduced by mischievous behavior of some individuals, or the hazards to unattended passengers egressing to the roadbed in the presence of a "live" third rail and trains on adjacent tracks.
3. The importance of continued operation of the electrical system cannot be overstressed. Wires carrying current for lights, loudspeakers, and control systems must continue to function so that doors can be opened and passengers can see means of egress and be directed in evacuation procedures. Electrical insulation must maintain its integrity even with direct flame impingement. Cable carrying current to the traction motors also should be able to resist direct flame impingement long enough to move the train to a safer place for the passengers to evacuate if deemed expedient.
3. Transit system crews must be trained thoroughly in emergency evacuation procedures and be required to rehearse those procedures at regular intervals.

7.1.4.4.2 Emergency Crew Hazards

The foremost responsibility of the operating personnel in the event of fire is the safe evacuation of the passengers. After passenger evacuation, they may attempt to extinguish the fire with whatever extinguishment equipment is at hand. The fire department must be notified at the first opportunity.

Smoke usually will be the first hazard encountered by firefighters; they must be suitably equipped with oxygen packs to enable them to enter the area, to find passengers who have not been able to escape, and to extinguish the fire. Depending on the type of operation, live wires, cables and a live third-rail or catenary would be additional hazards.

7.1.4.5 Current Fire Protection and Control Systems

7.1.4.5.1 Prevention

Fire prevention begins with the design and construction of the transit system. Unfortunately, the large majority of the transit systems are old and were designed under the concept of fire prevention prevalent at the time of construction. The designers of new systems should take advantage of contemporary technology to provide the most fire-resistant system possible.

The first step in fire prevention is to limit the amount of things that can burn. Since combustible materials are bound to be present, every effort should be made to:

1. Keep the quantity of combustible materials to the smallest amount, consistent with minimum comfort and operational requirements.
2. Use polymeric materials exhibiting the greatest resistance to ignition, fire spread, and smoke emission.

Commercial operations such as food stores in a transit facility are undesirable from a fire safety standpoint and should be avoided. If this is not possible, they should be separated from the transit operation by a minimum 2-hour fire rated enclosure.

Proper maintenance is another key to fire prevention. Again, the concept is the elimination of trash in the storage of combustibles in tunnels. Undercar components must be kept free of grease and metal particles by frequent cleaning. Tunnel vacuuming, which is practiced, as previously noted, in several modern systems, is an excellent preventive measure. Fire safety devices must be kept in good repair.

7.1.4.5.2 Detection

Generally, the detection of fires on rail systems is dependent on human observation, and this method works very well for occupied passenger compartments. However, fires involving the underside of a car often can burn undetected for a considerable period of time unless observed by the motorman of a passing train and the alarm given.

Smoke detectors now are considered to be as effective as human observers on vehicles where a responsible attendant can be summoned quickly. On completely automated vehicles, however, a smoke detector of the most reliable type is a basic requirement.

7.1.4.5.3 Maintenance of Egress Assistance Aids

Ease and speed of passenger egress are important to safety, and therefore, affect material selection and use. Hence, it appears that brief discussions of egress aids and methods are in order. A concept almost universally accepted by the transit industry is that if a train should be discovered to be on fire, the operator will make every effort to get that train to the nearest station. The one exception would be that if a train were on the surface, the operator would not enter a tunnel to reach a station. Once in the station, the train can be evacuated quickly and firefighters' job would be made easier because of improved access.

It is sometimes impossible for a train to reach a station in an emergency because of physical damage. In that case, a train will be stopped in a tunnel, or on a surface right-of-way or aerial structure, making evacuation extremely difficult. A fully loaded 8-car train could be carrying as many as 1,500 passengers, and panic may be an added factor that might impede orderly evacuation.

To accommodate the evacuation of a train not in a station, most transit systems provide some type of ladder to overcome the problem of the 40-inch "step" between the car floor and the track bed. These ladders can be anything from a specially built folding ladder designed to fit exactly into the door of a car to an ordinary step ladder.

Egress through the doors may be easy in many emergency situations. However, if the doors are inoperable because of power shut-off and the inability of passengers to follow the emergency opening procedures, the windows are the final means of egress. It has been mentioned previously that many vehicles now are fitted with plastic windows to reduce breakage caused by vandalism. If the plastic is polycarbonate, the windows may be exceedingly difficult to break; therefore, new cars using polycarbonate windows have special gaskets containing a removable strip of elastomer attached to a metal ring. When the ring is pulled, the strip, which goes completely around the window, is removed, and the window can easily be pushed out. This feature, when used, is not necessarily provided for all windows in a given rail vehicle.

7.1.4.6 Post-Fire Initiation Considerations

7.1.4.6.1 Fire Location and Type

The type of fire that results from a given ignition source is determined by the adjacent material as well as the type and magnitude of the fire load to support its development. Ignition sources may develop smoldering fires or flaming fires. Thus, a whole range of possibilities exist. As indicated above, the majority of fires in fixed guideway vehicles will originate under the car; therefore, the floor must be considered as a prime fire barrier. All means of conveying the fire to the interior of the passenger compartment must be designed to resist penetration. These means include electrical passthroughs and any ducting that may originate under the car.

7.1.4.6.2 Design, Operation, and Procedures for Handling Emergencies

Even though a modern rapid rail transit system is inherently fire safe and the fire safety record of the transit industry is excellent, contingency plans must be prepared to handle emergencies that may occur. It is perhaps just such emergency procedures that have allowed the transit industry to establish its enviable fire safety record.

Fire safety for any type of occupancy must start in the planning stage. Proper design of the system itself and the prudent installation of fire detection and suppression equipment is essential for a successful fire safety program. In designing new systems, the transit industry has chosen to build with fire-retardant materials and fire-rated walls separating unlike areas. This policy has resulted in the need for few or no automatic fire suppression devices. However, such equipment is installed where the hazard warrants. On the other hand, no matter what materials are used in construction, a fire is still possible, and smoke detectors should be installed in all non-public areas of a system. Provisions also must be made to monitor such detection equipment continuously.

Even the best fire protection will not prevent every fire. When a fire does occur, the difference between a minor and major fire will be determined by advance planning and personnel training. It cannot be emphasized too strongly that such preplanning and training must include the local public fire service. The transit authority can go only so far to control a fire emergency and the fire service must be thoroughly prepared to take over the incident and see it through to completion.

Reporting of emergencies is often a process in which valuable time is wasted. Transit authority employees who even suspect that there is a fire or other emergency involving transit property must immediately advise the control center and the control center must immediately notify the appropriate fire rescue service. In most jurisdictions this is all too often overlooked. Small fires can be extinguished easily, but delay in notification can result in a large fire.

If the fire is suspected to be in a station, the person assigned to that station can investigate the situation after notifying the control center of the suspicion of a fire. In the case of unattended stations or tunnels where automatic detection equipment has been activated, transit authority personnel can be dispatched to the station even though the fire department probably will arrive first.

The nature of the materials in the station, the location of the fire, and its severity are facts that will be used to decide whether the station would be evacuated. For example, activation of detection equipment in a service room with no visible smoke probably will not require evacuation. However, if a train enters the station with a car on fire, the need to evacuate is obvious.

When the need to evacuate a transit station arises, it should be done using a public address system rather than the ringing of a bell. People tend to ignore fire alarm bells if there is nothing to make them suspect there is any danger; however, a fire alarm bell coupled with an odor of smoke can cause severe panic. A calm commanding voice over a public address system can better direct passengers.

Responsible personnel have several options on how to advise passengers to evacuate a station depending on conditions. Generally, the means of exit will be the same as the means of entry via the normal exits. If the fire involves an escalator at the only entrance, passengers may be directed to board trains to be removed from the station, to use an emergency exit shaft if one exists, or, in an extreme emergency, to enter a tunnel and proceed to an emergency exit shaft or the next station (after the power in the third rail has been shut off).

Should a fire involve a train in revenue service, the operator should advise the control center of the situation. In no case should a train operator stop a train on fire in a tunnel; instead he should proceed to the next fire station where the passengers can be evacuated and firefighting operations begin. An exception to this rule would be that a train operating on a surface right-of-way should not enter a tunnel to reach a station.

If a train becomes disabled in a tunnel for whatever reason, evacuating passengers to the track bed and walking them to an emergency exit or a station will be used only as a last resort. Other methods of evacuating passengers involve having passengers go from the disabled car into serviceable car in the same train, cutting away the disabled car, and proceeding to the next station or bringing another train up to the disabled train and evacuating passengers into the rescue train. These are the two basic evacuation schemes, but there are several variations of each.

Automatic fire detection and suppression systems probably are not practical for transit vehicles considering the early state of the art. The high temperatures and air-suspended particulate matter found under the cars, in addition to the presence of potential vandals inside the car, tend to discourage the use of automatic detection and suppression equipment. One type of equipment that may be practical, however, would be a fixed fire extinguishing system that could be activated manually. It could be directed to handle spot fires in high hazard areas.

In most transit systems the local fire department is responsible for providing fire protection service. However, the transit system must provide an adequate number of fire extinguishers of the proper type to ensure that an employee discovering an incipient fire will have a high probability of being able to extinguish it. For more advanced fires, standpipes for fire department use should be provided in all stations and tunnels.

Fire occurring in tunnels can be expected to produce large quantities of dense black smoke. To assist with the evacuation of passengers and firefighting operations, a ventilation system in the tunnels is a necessity. The ideal system would encompass fan shafts in which reversible fans could be controlled from a central point. A manual control at the fan also would be needed should the central control fail. Reversible fans would allow the smoke to be directed to keep the paths of exit travel and the stations clear. A plan for fan operation must be established prior to an emergency.

7.1.4.6.3 System Operations Affected by Fire Safety Considerations

The movement of passengers in a transit system generally is not greatly affected by fire safety. However, as discussed above, fire safety is built into the system and protects passively. Probably the only fire safety consideration that could greatly affect passengers would be a prohibition against smoking on board rail transit vehicles.

Fire safety has the greatest impact on the daily operations of the maintenance department. Notably, there is a prohibition against the use of Class I flammable liquids or equipment fueled with Class I flammable liquids in any subsurface environment. This rule often delays work until the proper diesel, electric, or other type of power equipment can be obtained. Welding and most metal cutting operations also require special precautions such as the use of a fire extinguisher and the services of a watchman.

As part of their daily start-up procedures, all operations employees must inspect fire protection equipment in their areas of responsibility. Train operators must check the fire extinguishers on their trains and the emergency ladders. Station attendants also must check fire extinguishers, annunciator panels on the fire detection apparatus and emergency exit doors. In a modern system, the fire alarm would also be electronically monitored and an appropriate alert signal sent to the control center should a malfunction occur.

Two important effects of tunnel fires add more urgency to the requirements for more fire safe vehicles. It was demonstrated recently that the combustible materials in a vehicle can generate sufficient heat to melt an aluminum car shell and allow it to flow over the tracks. The tunnel then is effectively shut down until the car and tunnel cool and crews can cut up the vehicle for removal.

An intense fire in a tunnel also can affect the concrete tunnel liners and cause spalling and cracking of the concrete and, possibly, weakening of the structure to the point of requiring its rebuilding.

7.2 Materials Used in Transport Systems

For fire safety of fixed guideway systems, one must be concerned with all the components thereof. In addition to the vehicles, these systems contain stations, tunnels, guideways, vehicle storage areas, and control rooms.

Stations have not been directly involved in passenger fire hazards for the most part because of the limited amount of combustibles found in them. A notable exception to this general observation, however, are combustibles in stores and concessions found in many stations. In one case, there was a severe fire in a bakery concession located on the mezzanine level of a subway tunnel. Smoke from this fire filled the rail tunnel, making the transit lines inoperative for several hours. Stations should be free from this type of hazard because they are the vital parts of passenger evacuation routes.

Polymeric materials have been used as noise attenuators in stations in order to reduce the very high noise level when a train rolls into a station. Because of susceptibility to vandalism, however, polymers have been abandoned in favor of porous ceramics.

A fire in a subway tunnel is considered by transit officials to be their number one safety problem. The worst type of subway fire is one that might occur in a tunnel under a river or a bay. When the tunnel is located under the streets, there are exists to the streets at given intervals. A tunnel under a river or other body of water or a hill has only two exits — one at either end of the tunnel — that sometimes are miles apart, making escape from a hazardous situation especially difficult. (The situation is similar to that encountered in an air-borne aircraft.)

In those cases, the first line of protection against fire is the materials themselves. The materials must be the most resistant to ignition that can be specified and selected by trade-offs of risk versus cost. Smoke and toxic gas emissions must be kept to a minimum by specifications that prohibit the use of materials that produce copious amounts of smoke and toxic quantities of gases.

While not all vehicles use nonmetallic materials in all categories, the following are most common:

1. Wall and ceiling panels.
2. Seat frames and shrouds.
3. Seat cushions and fabrics.
4. Flooring.
5. Carpeting and floor tiles.
6. Plastic glazing.
7. Hydraulic lines and fluids.
8. Lighting diffusers.
9. Thermal and acoustical insulation.
10. Electrical insulation.
11. Vehicle shelves and end-caps.

12. Air-conditioning ducts.
13. Fuel tanks and fuel.
14. Tires.

Table 3 provides a tabulation of materials used in rapid transit and commuter cars. These parts and assemblies will be discussed further below.

7.2.1 Seat Cushions, Fabrics, and Frames

In vehicle interiors, arson is a major cause of fires, and seat cushions are the prime targets of the arsonist. Vehicles having reinforced plastic seats are much less vulnerable.

Fabric-covered polyurethane and neoprene foams are two types of material combinations in general use for seats. Even if highly flame-retardant fabrics are used for covers, the arsonist often comes equipped with a knife as well as matches or a lighter, and the foam interior is quickly exposed. Furthermore, some seats in public vehicles do not have metal pans under the seats, thus the foam is exposed so that ignition of a newspaper under a seat could readily ignite the foam. Since the flexible polyurethane foams have been found to be so readily ignitable, neoprene is being specified more and more for new vehicles and in the refurbishing of older vehicles.

Neoprene evolves dense black smoke and hydrogen chloride gas when ignited, but by the ASTM E162 Radiant Panel Test, neoprene has a flame spread latex (I_s) of less than 10. The flexible polyurethane foams, on the other hand, do not exhibit an I_s of less than 300, and many are considerably in excess of this number. When one is buying time for the evacuation of passengers and to permit the approach of emergency personnel, the material that is more difficult to ignite and that does not immediately involve adjacent materials is preferred.

Tests by the National Bureau of Standards (1975) on a bus and a simulated rail rapid transit car of the Washington Metro System have indicated that:

1. Within 1 or 2 minutes after the ignition of a single polyurethane cushion in the bus, visibility was essentially zero.
2. Polyurethane seat cushions involve the rest of a vehicle in fire much more quickly than neoprene foam cushions.

Three promising new seat cushion foam materials are now under development; these are polyphosphazenes, silicone sponge, and polyimides.

The seat cushion fabric that has been in most general use in public transportation vehicles is polyvinyl chloride-coated fabric. It is a very durable and easily maintainable material and is quite resistant to fire, as shown by FAR 25.853 Vertical Test. However, on exposure to heat, or on ignition, it releases high-density smoke and hydrogen chloride gas. Other fabrics in use that do meet FAR flammability regulation and proposed smoke regulation are wool, Nomex®, and wool/nylon blend. Wool however, is a known emitter of hydrogen cyanide. Fire safe fabrics are high on the list of desired materials.

Seat frames that are not metal usually are made of fiber-reinforced plastic. These should be required to meet the same flammability and smoke standards as the wall and ceiling panels.

Table 3. Tabulation of Materials Used in Various Rapid Transit and Commuter Cars.

Transit System	A	B	C	D	E	F	G	H	I	J	K	L	M	N	O
Body Structure															
Ends	FGPR SS	FGPR SS	SS	FGPR	SS	Al	SS	SS	FGPR SS	FGPR	Steel	LAHT	FGPR	Al	SS
Sides	SS	SS	SS	Al	SS	Al	SS	SS	SS	Al	Steel	LAHT	SS	Al	SS
Roof	SS	SS	SS	Al	SS	Al	SS	SS	SS	Al	Steel	LAHT	SS	Al	SS
Floor	SS	SS	SS	Al	SS	LAHT	SS	SS	SS	Al	Steel	LAHT	SS	Al	SS
End under frame	LAHT	LAHT	LAHT	LAHT	LAHT	LAHT	LAHT	LAHT	LAHT	Al	Steel	LAHT	LAHT	Al	LAHT
Inside Linings															
Side	KRoyNor	Roy	MLP	ACR-PVC	MB	MLP	MLP	MLP	NyCar on Al	ABS	Al/ MAS	MAL	MPM	MAL	P/AL
Side Ceiling	MB	MB	MAL	ACR-PVC	MB	MB	MAL	MB	NyCar & MLP	FGPR	Al	MPM	MB	MAL	P/AL
Ceiling center	VAL	VAL	MNPM	MHC	MB	MPM	MPM	MNPM	MPM	FGPR	COMP	MPM	MB	MAL	PM
End	MPM	MPM	MPM	ACR-PVC	MLP	MPM	MPM	MPM	MPM	ABS	Al	MPM	MPM	MAL	P/AL
Window masks	AL,RBR	AL,RBR	Kydex	ACR-PVC	Kydex	ABS	FGPR	ABS	FGPR	---	BRASS	Al	---	FGPR	P/AL
Partitions	MPM	MPM	MPM	MPM	MPM	MPM	MPM	MPM	MPM	---	Al	MPM	MPM	MAL	PM
Floor															
Sheets	3/4AHPM	3/4APM	3/4APM	5/8APM	3/4SPMI	3/4APM	3/4APM	3/4SHPM	3/4APM	Al-UR	StTrPl	3/4GSBP	3/4RPM	3/4PW	1/2PW
Covering	RBR Ti WoCar	RBR Ti WoCar	ShRBR	WoCar	ShRBR	ShRBR WoCar	RBR Ti	ShRBR	NyCar RBR Ti	WoCar	Con.	ShRBR	ShRBR	RBR Ti	RBR Ti
Insulation															
Sides	FBGLS	FBGLS	FBGLS	FBGLS	RgdUR	FBGLS	EmUR	FBGLS	FBGLS	FmUR	HRFLT	FBGLS	FBGLS	FmUR	FBGLS
Roof	FBGLS	FBGLS	FBGLS	FBGLS	RgdUR	FBGLS	FmUR	FBGLS	FBGLS	FmUR	HRFLT	FBGLS	FBGLS	FmUR	FBGLS
Floor	FBGLS	FBGLS	FBGLS	FBGLS	RgdUR	FBGLS	FBGLS	FBGLS	FBGLS	FmUR	None	FBGLS	FBGLS	FmUR	FBGLS
Seats															
Cushions	NEO	NEO	UR	NEO	SPRNGS UR	NEO	NEO	NEO	NEO	UR	SPRNGS HRFLT	UR	SPRNGS	UR	UR
Covers	VINYL COTTON	VINYL COTTON	VINYL	VINYL	VINYL	VINYL WOOL	VINYL WOOL	VINYL	WOOL	VINYL WOOL	RATH/ CANV	VINYL	VINYL DUCK	VINYL FABRIC	VINYL
Frames	STEEL	STEEL	STEEL PW	STEEL	STEEL PW	STEEL PW	STEEL PW	STEEL PW	STEEL	STEEL	Al	METAL	SS	STEEL	STEEL
Supports	ShRBR	ShRBR	RBR Straps	ACR-PVC	NONE	ShRBR	RBR Straps	ShRBR	ShRBR	---	STEEL	ShRBR	NONE	FBGLS	PW
Back	KYDEX ROY	ROY	ROY	ROY?	SS	WO VIN	KYDEX	WO VIN	WOOL FABRIC	KYDEX	---	ROY	ROY	SS	WO WIN
Light Fixture															
Lens	PC	PC	ACR	PC	ACR	ACR	ACR	PC	ACR GLASS	ACR	NONE	AL EGG Crate	GLASS	ACR	ACR
Wire Insulation	HYP TEF	HYP TEF	RBR NEO	HYP TEF	TEF	HYP	RBR NEO	HYP TEF	RBR NEO	TEF	RBR COT,BRD	HYP	RBR PVC	NEO RBR	HYP

LEGEND:

ABS—ABS-type Thermoplastic
ACR—Acrylic
3/4 AHPM—Al PM with honeycomb core (3/4" thick)
Al, AL—Aluminum
3/4 APM—Al PM 3/4" thick
COMP—Composition board
CON—Concrete
COT BRD—Cotton braid
FBGLS—Fiberglass
FGPR—Rigid Fiberglass
3/4 GSBP—3/4" Galvanized Steel Faced Plywood
HRFLT—Hairflet
HYP—Hypalon
K—Kydex
LAHT—Low Alloy High Tensile Steel
MAL—Melamine on Al
MB—Balanced Melamine
MHC—Melamine on Al-faced honeycomb
MLP—Melamine on Laminated Plastic
MNPM—Melamine on Novaplymetal
MPM—Melamine on Plymetal
NEO—Neoprene
NOR—Noryl
NyCar—Nylon Carpet
P/AL—Painted Aluminum
PC—Polycarbonate
PM—Plymetal
PVC—Polyvinyl Chloride
PW—Plywood
3/4 PW—3/4" Plywood
RATN/CANV—Rattan on Canvas
RBR—Rubber (Ti-tile, Sh-sheet)
ROY—Royalite
3/4 SHPM—3/4" Stainless Steel Faced honeycomb Plymetal
3/4 SPMI—3/4" Plywood with Stainless Steel on Top Face
SS—Stainless Steel
TEF—Teflon
UR—Urethane (Fm-foamed, Rgd-rigid)
VAL—Vinyl on Aluminum
Wo Car—Wool Carpet
Wo VIN—Vinyl on walk-over type seat

7.2.2 Wall and Ceiling Panels

Early vehicles had wooden interior wall and ceiling panels, which were eventually replaced with painted or waffled steel or aluminum. The first plastic material that became available for this application was based on a melamine-formaldehyde resin.

It was used extensively. Some melamine resin formulations have good fire-retardant and smoke-emission properties; however, melamine resin requires more expensive match-metal die molds than the inexpensive thermoforming materials, and, hence, has lost favor with cost-oriented designers of modern cars. As a result, panels of polyvinyl chloride/polymethyl methacrylate (PVC/PMMA) and polyvinyl chloride/acrylonitrile-butadiene-styrene (PVC/ABS) now are widely used.

PVC blends have been unable to meet new smoke specifications. Consequently, there is now much activity to develop new materials, such as highly filled polycarbonates, in addition to fiberglass/reinforced polyesters, epoxies, furanes, and polysulfones. From the standpoint of flammability and smoke, a preferred material would be metal panels with a thin polymer film on their surfaces to provide color and aesthetic finish.

There also has been a recent trend to carpet at least part of the walls and even the ceilings of some vehicles with a concomitant increase in the fire hazard. Such applications are nonessential.

7.2.3 Flooring and Fire Walls

Vehicles that carry fuel beneath the floor or that have high-voltage cable connections, or the third-rail power pick-up located beneath the floor must treat the floor as a prime fire barrier. In a study of electrical fires in transit systems, it was found that approximately 75 percent of all vehicle fires originate in electrical short circuits of various kinds in components beneath the floor. Thus, the floor fire barrier must provide sufficient time for the passengers to escape from the vehicle.

In the drive towards reducing the weight of vehicles as a means of conserving energy or reducing cost, the fire resistance of the flooring, along with other vehicle parts has been adversely affected. For example, floors in certain rapid transit cars have been constructed of polyurethane foam sandwiches with 1/16-inch aluminum sheets on each side to reduce weight. With this type of construction, an under-car fire can readily melt the aluminum and burn through. Floors and walls also must keep smoke from a smoldering or flaming fire out of the passenger compartment. It is therefore necessary to design so that pass-throughs for wires and air-conditioning ducts do not become the means for fire and smoke penetration. An adequate floor construction used in many vehicles consists of a 3/4-inch fire-retardant treated plywood faced on each side with 25-gauge (.025") sheet steel.

Since there are no specific tests for floors, it is suggested here that flooring or fire wall test procedures be developed. The test apparatus should be of an oil burner type. The floor should be tested from underneath. It should be required to withstand penetration for a specified time (consonant with safe egress of all passengers from the vehicle).

7.2.4 Floor Covering

Floor coverings in vehicles now are made either of rubber or vinyl composition tile or of various types of natural and synthetic fiber carpeting. Actual experience in fires has shown no special hazard associated with the tile. Carpeting has become a popular floor covering in many types of mass transportation vehicles, but as noted above, some designers have even used floor carpeting on the walls and ceilings, thereby increasing the fire hazard.

The principal flammability test for carpeting in vehicles in recent years has been the ASTM E84 Tunnel Test. However, no specific limits were set. This test has now been determined to be an unacceptable testing standard for carpets normally used on floors. The Port Authority of New York and New Jersey adopted the ASTM E162 Radiant Panel Test with various limits for the flame propagation index of from 75 to 200 depending on the location of the carpet as the appropriate test for carpeting in vehicles.

The extensive studies of carpet flammability conducted by the National Bureau of Standards in its corridor tests and in the development of new Flooring Radiant Panel Test, have demonstrated the very wide range of flammability in carpeting. Data based on the Flooring Radiant Panel Test shows that carpets can ignite and continue to burn on exposure to a critical radiant flux from less than 0.1 watt/cm^2 for polypropylene to over 1 watt/cm^2 for some wool carpeting. Carpeting should be tested with any underlay that is to be used with it.

7.2.5 Plastic Glazing and Lighting Diffusers

Driven primarily by the high rate of vandalism, vehicle manufacturers have been turning to plastic windows. The two types being used are polycarbonate and polymethyl methacrylate. Stretched acrylic windows have been in use on aircraft for many years, and the use of plastic windows in buses in one large city has reduced window replacement from 50 percent to about 1 percent per year.

In some cases, a single sheet of the plastic is used and in others, double glazing is used in which the plastic pane is on the outside and safety glass is on the inside with an air gap in between. Both types of plastic window have mar-resistant coatings that help to extend their useful life.

Although both polymethyl methacrylate and polycarbonate are superior to glass in terms of impact resistance; polycarbonates are distinctly superior in fire resistance at a cost increase of about fourfold for the material. The Izod impact strength of polycarbonates is 16 whereas that for acrylics is about 0.5 (i.e., polycarbonate is about 32 times more resistant to impact). The flammability of polycarbonate is distinctly less than that of the acrylics (e.g., the flame spread index is 100 for the polycarbonate and 320 for the acrylic). Corner fire tests also demonstrate the superiority of the polycarbonates.

Adoption of polycarbonates indicates the value of fire scenario analysis. In this case, the polycarbonate is so impact resistant that firemen would have extreme difficulty entering through the windows, which they traditionally break. Some fire departments now regularly carry a special rotary saw since their axes are unable to penetrate polycarbonate windows readily. In some vehicles, special gasketing has been developed to enable passengers to push out an entire window by pulling a cord that removes a strip of elastomer set into the elastomeric gasket. Similar devices are set on the outside to permit access from the outside of the vehicle.

Lighting diffusers in many transit vehicles, at one time consisted of individual glass globes. They now generally consist of plastic covers over fluorescent tubes and run the length of the vehicle. Plastics have been found to be functional for this purpose; a typical lighting diffuser consists of a plastic strip about 10 inches wide running the length of the vehicle. This application can serve as an avenue for rapidly conducting a fire the full length of a vehicle.

In addition, some vehicles have begun to carry illuminated advertising in broad strips just above the windows on each side of the vehicle. In one such vehicle, the plastic covering consisted of polystyrene sheets. When this vehicle caught fire, the polystyrene was a significant contributor to the heat load.

7.2.6 Thermal and Acoustical Insulation

A single material usually is used to perform the functions of both a thermal and acoustical insulator. It fills the spaces between the inner and outer walls and ceiling as well as serving as the floor sandwich described earlier.

One material that is an excellent thermal insulator, is easy to apply, and is inexpensive is sprayed-on polyurethane foam. However, as noted above, polyurethane foam is relatively easy to ignite and, upon burning, contributes significantly to the total heat load. Indeed, the wall spaces in which the insulation is placed sometimes are used as a plenum for the air-conditioning system and, thus, are potentially a means for carrying fire and smoke to all parts of the vehicle. Polyurethane foam also has been used to insulate engine compartments in various vehicles, but this application has now been dropped because of a high incidence of fires.

Many of the new mass transportation vehicles now feature fiberglass batts for insulation. If not more than 2 to 3 percent of resin is used as a binder, this material performs its functions very well and eliminates a potential fire hazard. Mineral fibers also will be satisfactory if they are installed so they will not sag with the motion and vibration of the vehicle.

7.2.7 Vehicle Shells and End-Caps

An attempt to reduce the weight of vehicles and to lower costs has resulted in designs that call for molded plastics for the shells of rail rapid transit vehicles and personal rapid transit vehicles (PRT). Approximately 75 percent of all fires experienced in this type of system come from undercar electrical sources, and such a fire would quickly involve the vehicle shell. Not only would a fire involving a shell make it difficult to evacuate passengers, it also would make the approach of emergency personnel very difficult.

Although plastic end-caps use much less combustible material, their use also must be questioned. Recent full-scale tests have indicated that they would be a means of spreading a fire from car to car, especially when the train is in motion.

In considering the rate of flame spread and the localization of the heat load, one must consider the radiative effects from the walls of a tunnel. Actual tunnel fires, such as the one that occurred in the Boston street car tunnel during July 1975, have revealed how a fire in one vehicle can cause radiated heat from the tunnel walls to ignite an attached car.

The use of plastics in the shells also should be considered in terms of property damage. A simple paper fire on a platform along side a current plastic shelled vehicle could cause involvement of the entire vehicle.

7.2.8 Electrical Insulation

Electrical short circuits and current overloads immediately affect polymeric electrical insulation causing it to smolder or to burst into flames. The smoke arising from this source is also a major concern. Transit authorities have long sought a "smokeless cable" for use in their vehicles as well as in wayside installations.

Maintaining the integrity of electrical circuits during a fire also is of great importance. Traction power is needed in some cases to move the vehicle to a better position for evacuation of passengers. Continuation of lighting, control, and public address systems are vital in a vehicle tunnel fire since total darkness and lack of direction can lead to panic. Consequently, insulation is required that will emit little smoke, and retain its insulation characteristics when subjected to the high temperatures of direct flame. Some electrical insulation materials now in use can resist high temperatures, but they eventually melt or fall off the wire, resulting in short circuits. There are, however, some insulations that remain on the wire even after being subjected to direct flame. Insulations that retain their integrity can be developed by using mica flake, asbestos, or other inorganic materials as one of the insulating layers.

A current program, funded by the Department of Transportation, is investigating flammability, smoke emission, and toxicity characteristics of burning electrical insulation from the smallest wire size to the large cable used for traction motors. This program will determine the best test methods to evaluate flammability and how best to use the NBS Smoke Chamber (NFPA 258) for measuring smoke emission. Toxicity of various insulations at several different exposure temperatures will be determined using the live rat method developed by the FAA Civil Aeromedical Institute.

7.2.9 Hydraulic Lines and Fluids

By virtue of their design function, hydraulic lines often run through locations where the hazard of fire is greatest. The lines usually are run under a vehicle or through an engine compartment. Hydraulic fluid is stored in a reservoir that may contain up to several gallons and is distributed by pumps under pressure to the various functional parts.

Where hydraulic fluid is used to activate brakes, the line used to feed it into the brake cylinder is usually flexible and often consists of neoprene or a fiber-reinforced elastomer. If this line should break or be burned by brake malfunctions, hydraulic fluid can be sprayed into the fire and contribute substantially to the heat load.

Many types of hydraulic fluids are available. They range from highly combustible hydrocarbon oils to synthetic fluids that are considerably more fire retardant. The U.S. Air Force and Navy have developed hydraulic fluids that can provide a greater measure of fire safety than those now in use in many transit systems.

7.3 Tests

Test methods, specifications, and standards are considered in detail in Chapter 5. An extensive review of flammability tests is presented in Volume 2.

7.4 Smoke and Toxicity

Smoke emission test procedures are described in Section 5.4.6, and toxicity is discussed in Section 5.4.7 as well as in Volume 3.

7.5 New Concepts and Developments

The tightening of flammability and smoke specifications for materials for use in transit vehicles has resulted in a fresh look at phenolic resins and consideration of some newer materials such as furanes, polysulfones, bismaleimides, polyphosphazenes, and polybenzimidazoles. Flame-retardant additives have been evaluated extensively. Some of these substances such as antimony oxide, are found to cause increased smoke emission. Recently, alumina trihydrate has become a preferred filler for sheet plastics. It can be used at very high loadings in reinforced plastic with due regard to its adverse effect on strength. This practice has even improved the properties of some polyesters to the extent that they have met the more stringent safety requirements.

New concepts are being developed to incorporate the rate of heat release and the total heat release of materials into specifications that now determine only ease of ignition and rate of flame spread. Another concept in development is the "Total Hazard Index." It is hoped that the method will permit the combination of measures of flammability, smoke emission, and toxicity in an equation to provide a figure of merit for a single material or for a combination of materials.

A useful toxicity standard appears to be nearing realization in work being done by the FAA Civil Aeromedical Institute and others.

7.6 Suppression Systems

Fire suppression systems for transit systems affect public safety, particularly in regard to successful egress, and, thus, affect materials choice and use. They consist of several categories: (a) wet and dry standpipes established in tunnels to which hoses are attached in the event of tunnel fires, (b) water pipes with spray nozzles laid down in stations between tracks, (c) hand-operated fire extinguishers provided in stations and on cars, and (d) automatic Halon 1301 systems for use on small, unattended automated vehicles. Fire suppression systems are excluded from the committee's major interests and are not discussed or analyzed further.

7.7 Special Considerations for Unattended Vehicles

New vehicles are being designed and used that are totally automated, have no transit personnel on board, and whose performance is monitored at a central control station. A few of these systems are now in operation at airports and for other short runs. The vehicles operate on fixed guideways and function in several different modes; e.g., some are rubber-tired on concrete guideways, some are air-cushioned, and some are suspended on monorails.

The suspended monorails introduce great difficulty in a fire. Since the vehicle may be suspended anywhere from 10 to 30 feet above the ground or over water, the most logical means of egress are rope ladders or fabric chutes, but occupants of such vehicles might be young children or elderly or handicapped people who are unable to use these evacuation methods. An alternative to evacuation is to keep the passengers in the vehicles until they can be removed or until the vehicle can be pushed to the next station. For this procedure to be feasible, an automatic fire-extinguishing system must be activated by a smoke detector. Moreover, the extinguishant must extinguish the fire immediately without it self providing a toxic hazard for the occupants of the car. Under such circumstances, Halon 1301 (CF_3Br) provides such protection. However, this extinguishant can be used only in the presence of people if the fire is a "surface" type capable of instant extinguishment. A deepseated fire will allow thermal decomposition of the CF_3Br with the production of toxic substances.

It is evident that completely automated vehicles have all the problems of attended vehicles as well as additional problems incurred by the absence of persons trained to handle emergencies. It is therefore necessary that the materials of construction used in these vehicles be as fire safe as possible, that a minimum amount of combustible material be used compatible with comfort, that a smoke detector be used that will have the highest possible reliability, and that passenger smoking be prohibited.

7.8 Special Military Use Considerations

With regard to fixed guideway passenger vehicles, there are no special military requirements beyond those stated above.

7.9 Conclusions and Recommendations

Conclusion: Fixed guideway vehicles operate under conditions that can rapidly develop into hazardous situations. Many of the polymeric materials currently used in ground mass transportation vehicles are potentially hazardous in that they reduce the effectiveness of fire protection in the transit systems and contribute to increased fire hazard from the production of smoke and toxic gases. *Recommendation:* Use only those polymeric materials that, by testing and comparison, are judged to be the most fire retardant and have the lowest smoke and toxic gas emission rates; even these materials should be used sparingly, consistent with comfort and serviceability.

Conclusion: Undue consideration has been given to aesthetics and comfort in recent car designs and construction. Certain polymers have been improperly and excessively used; this has resulted in unnecessary fire loads and increased fire susceptibility. *Recommendation:* Replace the PVC formulations currently in wide use as wall and ceiling panels with more flame-retardant, less smoke-emitting plastic panels; consideration also should be given to the use of metal panels with plastic coatings to provide desired colors and texture.

Conclusion: Because mass transit vehicles operate in tunnels and other hazardous areas, the materials of vehicle construction must be the most resistant to ignition, consistent with the necessary trade-off of risks versus cost. *Recommendation:* Design vehicles and transit systems using materials with the greatest potential for minimizing the threat of fire.

Conclusion: The floor of a fixed guideway passenger vehicle is the prime fire barrier between the passenger compartment and the operating components beneath the car where most fires originate. *Recommendation:* Require polymeric materials used in floor construction to provide the same resistance to penetration by fire as current plymetal (metal-wood composite) construction.

Conclusion: Continued operability of a passenger vehicle electrical system under fire conditions is a most important factor in total fire safety. *Recommendation:* Implement the recommendations of the several studies of the "smokeless cable" concept to provide protection from the fire hazards of electrical insulation and to ensure continuity of electrical service during fires.

Conclusion: Vehicle designers, manufacturers, and operators have inadequate knowledge of the fire-related properties of the polymeric materials being used. *Recommendation:* Make available automatically operated fire extinguishment devices in those situations where trained personnel are not in attendance as well as other fire extinguishment resources at frequent intervals at the wayside. Develop a methodology to determine the overall fire threat and fire hazards of a given system and to provide the data that will be used during the design phase to mitigate the fire threat.

Conclusion: Radiation of heat from the walls and ceiling of tunnels has been largely ignored in calculating the effects of heat loads on the fire safety of materials used in vehicle construction. *Recommendation:* Take into account the essentially closed system of a tunnel in making calculations of total heat load imparted by a burning vehicle. Fire-harden and provide the air-conditioning plenum in passenger vehicles with closures having fail-safe fusible links.

Conclusion: The use of presently known flexible polyurethane foam in seat cushions is not consistent with overall fire safety; polychloroprene (neoprene) foams currently are the only reasonable substitute cushion material. *Recommendation:* Do not use polyurethane foam in seat cushions; instead use polychloroprene foam.

CHAPTER 8

BUSES

8.1 System Design, Operation, and Fire Safety

8.1.1 Introduction

This chapter considers buses, trolley buses, and school buses propelled by internal combustion engines. The use of buses has grown rapidly during the twentieth century to the top place in the mass transportation industry. During the early years, materials used in the frame, body, and engine of buses closely paralleled those used in trucks and automobiles and materials used in bus interiors paralleled those used in street cars. Polymers, except wood, were used sparingly or not at all for interior furnishings and wall construction; polymers were used only for hoses, gaskets, belts, and tires in the engine and body. Ignition of interiors was extremely difficult, and ignition of exteriors was virtually impossible (except where fuel escaped and was ignited). Fire loads were small and well separated by steel and other metals. The few fires that occurred were, in the main, easily contained and extinguished.

Under the varying economic, political, social, and technological pressures of the 1950s, 1960s and 1970s, bus construction has changed dramatically. Polymers have replaced metal for many purposes, with substantial benefits to the bus passenger, bus operator, bus manufacturer, and society in general. These benefits have been accompanied by a sharp increase in fire susceptibility and fire load as well as a reduction in fire containment capability and ease of extinguishment. Dangers from smoke and toxicity have sharply increased. The fire safety aspect of increased polymer use in bus construction has been inadequately considered by specification writers, designers, manufacturers, operators, and passengers.

Fire statistics concerning buses (See Table 1) are relatively incomplete and somewhat misleading. The increase in fires reported does not indicate the sharp increase in danger to passengers (and the heavy monetary losses).

Not all fires in buses are catastrophic resulting in total destruction; not all polymeric materials as used, contribute to excessive fire load or fire susceptibility. As more polymers are used, however, the vulnerability will increase at a geometric rate unless fire safety becomes an integral part of the design process. Several photographs of bus fire results are presented in Figures 1-3 to illustrate the hazards involved.

The National Transportation Safety Board (NTSB) Safety Recommendations H-75-12 and H-75-13 of June 9, 1975, are quoted here in part:

> ". . .Though few in number, accidents involving fire are not only much more severe and costly than the average bus accident, but the number of fatalities is thirty times the number per accident for all bus accidents. . .

Table 1. Bus Fire Statistics.

	1973	1974
Systems Reporting	126	130*
Systems Reporting Fires	72	84
Number of Fires Reported	137	234
Systems Reporting Injuries	0	4
Number of Buses Destroyed	5	5
Reported Damage Value (Dollars)	126,959	137,095
Fire Locations and Causes		
Engine Compartment		
Electrical short	33	48
Fuel line break or leak	0	8
Defective exhaust system	4	3
Fires in carburetors	6	3
Miscellaneous and unknown	6	3
Wheel Wells		
Overheat of brakes	8	14
Flat tires	3	3
Tire tread wrapping around brakes	0	1
Power steering line failure	0	1
Electrical short	0	1
Passenger Compartment		
Vandalism	34	36
Electrical short	5	0
Lighted cigar on dash board	1	0
Other locations and nonspecified	10	17
Fires reported by groups, no locations	156	150
Fires reported with no causes	20	21
Total Fires Reported	281	314

Note: The total number of fires reported is greater than the sum of the items listed because of the many different ways of reporting and the voluntary nature of reporting.
*About 27,500 buses involved.

8.1.2 System Design Considerations from a Fire Safety Standpoint

8.1.2.1 Introduction

The bus, like most other forms of mass transportation, is not a completely self-contained unit (as is an airplane). It depends on its ambient (i.e., highways, surrounding environment, etc.). Under normal circumstances, bus fires should be survivable for physically able passengers since distance to egress points is shot. There are, however, serious potential survival and injury problems for aged or handicapped passengers and secondary risk problems for by-standers (e.g., on downtown streets where smoke and toxicity may incapacitate or when the fire occurs in a tunnel).

The fire load levels present in modern buses and the increasing numbers and types of fire ignition sources dictate a need to:

Figure 1. Bus fire results.

1. Define the fire hazard and fire control capabilities.
2. Reduce, where possible, the fire hazards.
3. Educate the American public concerning current risks.

Other than stated adherence to MVSS 302 and provisions of required firefighting equipment, the committee was unable to discern any manufacturers' design requirements relative to fire safety. Laudable industry efforts to improve safety and effectiveness of buses (such as the "Conventional School Bus Design Objectives" by the School Bus Manufacturers Institute, January 1, 1973) are silent on fire safety objectives. The new UMTA "Guidelines for Flammability and Smoke Emission," if adopted, will result in major improvement.

8.1.2.2 Considerations for Materials Used in Components

The functions of various components of bus construction usually are included in specifications and designs; but, performance in a fire situation is rarely included.

8.1.2.2.1 Part Function, Geometry, and Location

The design process usually involves a series of trade-off studies, many times resulting in compromises among requirements. If part function cannot be provided, other features become academic. With function established, the mandatory safety features, including fire aspects, can be established and the trade-offs initiated.

Figure 2. Bus fire results.

The geometry of bus interiors (seats, carpeting, window molding, luggage racks, side paneling, roof paneling, compartment panels, etc.) has a very important effect on flashover, particularly with regard to retaining heated gases and focusing explosively

Figure 3. Bus fire results.

ignited gases. Flame propagation rates vary according to the path of the flame front. While this aspect of geometry is extremely difficult to control, it should be considered, particularly in connection with ventilation system arrangements.

The location of a part can constitute a parameter for material selection by virtue of the response of the available polymers to the elements of selected fire scenarios. A material's propensity to drip, emit smoke and toxic gases, and self-extinguish and other fire performance characteristics can be extremely important. Materials that melt and drip would be hazardous if located so that the drops might fall on passengers or become additional ignition sources. Moreover, location of a part relative to possible elevated temperature and high rate of energy input as well as environment and its heat loss are important parameters for its ignition and sustained burning. Most materials burn faster and longer as the temperature of the environment is increased; some materials that are self-extinguishing (and meet specifications) at room temperature burn in a self-sustaining manner when the ambient temperature is as low as 250 °F (121 °C).

None of these factors are considered in relation to current bus material specifications. Clearly, designers, manufacturers, and bus purchasers must do so.

8.1.2.3 Part Versus Assembly

Proper evaluation of the combustion characteristics in a given application requires consideration of the material in combination with other materials in an assembly and the materials in adjacent assemblies.

Surface texture, color, shape, weave, gauge, density, heat content, etc., affect the ignition, burn, and smoke characteristics of otherwise chemically identical materials tested by the same method. Detail parts cannot be averaged to obtain their combined characteristics but must be tested as complete assemblies.

8.1.2.4 Assembly Versus System

Each assembly ultimately becomes a part of, or operates in, a bus system. Its fire performance may be considerably altered by system conditions. Assemblies in or part of the bus ventilation system normally will be subjected to a constant flow of air that will provide good support of combustion if ignition is obtained. Assemblies near exhaust system components normally will be exposed to elevated temperatures. Thus, assemblies that are in themselves satisfactorily fire retardant may not be satisfactory for use by virtue of system considerations.

8.1.3 System Operations as Affected by Fire Safety

The bus may be considered as a system as well as a portion of a super-system (i.e., a mass transportation system operating in a complex social environment).

The *ventilation subsystem* of a bus has important effects on fire safety by virtue of the materials of construction but more importantly by virtue of its operation during fire situations. Design and operation of bus ventilation systems have important effects on the fire dynamics involved in a bus fire; the fire dynamics are determined by the polymer species and the distribution of heat, smoke, and toxic gases in the bus.

Air usually is sucked into the bus by a fan, passed through filters, mixed with recirculated air (sometimes no recirculation is used), and passed through a ducting system or through an overhead plenum to distribution points in the bus. Exhaust openings collect air and pass it through ducting and exhaust fans to a discharge opening to the outside ambient. Ducting and plenums in older systems generally were of metallic construction, but new systems are increasingly being designed or modified to have polymeric materials in lieu of metals; such polymers have a wide range of compositions, combinations, and finishes.

Few ventilating systems have been designed with fire safety concepts adequately considered; most have no emergency closures or special operating procedures in a fire situation. Pending detailed analyses of selected fire scenarios, it appears that ventilating systems should be designed with fail-safe closures in the ducting at intervals (perhaps every 10 feet) and should be shut down in fire situations to avoid supplying additional air to the fire, spreading the fire by transmission of hot gases through ducting reducing the discharge of smoke and toxic gases to the atmosphere (particularly hazardous in downtown bus operating areas). New, large articulated buses particularly need this type of ventilation analysis.

Braking subsystens in operation, particularly at high speed, generate substantial heat and occasionally burst into flame. If these systems are not screened from passenger compartments by a fire wall, flames may penetrate. Many buses in operation today have fiberglass-reinforced plastic wheel well covers that can, and some have, catch fire from burning brakes (and tires) and permitted the flames to ignite the bus interior.

Electrical short circuits during operations produce fires that are easily transmitted in interor spaces where wiring or components are not protected by metal conduit or covers; continuing arcing provides a continuous ignition source for even those materials that are difficult to ignite. The trolley bus, with a continuous heavy electrical current flow possible at various locations at the top, requires special consideration during operations.

Fuel, while not a part of this study, clearly is a major fire consideration. Conditions during operations may be expected to cause serious magnifications of hazards if not properly considered, evaluated, and protected against during design.

Emission control and exhaust systems, with high exterior temperatures, pose special problems of material choice for nearby components, particularly during "stand-still" when normal convective cooling from bus movement is not available.

Power systems, hydraulic and air, not only have their own fire susceptibility problems but they also can affect nearby components.

While bus operations affect the fire safety of bus passengers dramatically, bus operation in the streets of a city may affect many others — on a crowded downtown street or in a tunnel — not only by flame, but also by smoke and toxicity. Thus, society's need and use of the bus must be considered critically in selecting materials for its construction.

8.1.4 Fire Scenaros as a Tool for Design Review

As discussed in some detail in Chapter 3 (and in greater detail in Volume 4), fire scenarios are beneficial not only as an aid in analyzing specific accidents, but also as a guide to bus designers in material selection and design modifications. In addition, they are helpful in developing realistic fire test methods and standards. Unfortunately, the current tests and standards for buses were not developed within the framework of performance in realistic fire conditions. Fire scenarios can guide the formulation of regulations and better design.

Scenarios have maximum utility if they: (1) represent bus fires having a high probability of significant injury or loss, and (2) provide sufficiently detailed information to permit useful analysis. Most bus fires are handicapped by the absence of trained observers, particularly during early stages, so analysts frequently must hypothesize from fragmentary evidence. Further, bus fire reporting lacks the cohesive approach necessary to produce sufficient statistical data to be of substantial use.

Under such circumstances, scenario development and use by designers is the best and perhaps only way currently available to provide a basis for optimum material selection in the fire situation.

8.1.4.1 A Fire Scenario — A Bus Fire Downtown

A nearly new bus had just started a run from its downtown terminus when a passenger told the bus operator that there was a fire at the rear of the bus (about 9:40 a.m.). The bus operator stopped the bus, went to the rear, and observed smoke coming from the roof as well as from a seat at the right side. He returned to the operator's station and attempted to call his base by radio but could not make contact. Meanwhile, the smoke intensified drastically and the few passengers on board hastily got off. The operator then got off the bus (9:42 a.m.). A passing bus operator notified the home base (9:44 a.m.); the first fire truck arrived about 9:45 a.m. and found that flames already had engulfed the bus and were shooting out of the roof. Firemen tried to put out the fire with water from the truck. This was not successful so the firemen hooked up to a fire hydrant. Flames ignited a nearby store awning, and radiant heat from the burning bus broke windows on the first two floors of the store. By 9:50 a.m. the entire interior of the bus was gutted with only the metal framing of the seats and the structure of the floor intact. All the seat cushions and covering, side interior paneling, wiring, plastic accessories, and the roof were totally burned. The store exterior was very hot, but early efforts by firemen prevented major damage to the store interior. The fire finally was completely out about 10:30 a.m.

An analysis of this fire scenario revealed that:

1. The bus had been built in full compliance with all specifications including MVSS 302.
2. The bus was well constructed.
3. Operations were normal and all equipment operating except for the bus radio transmitter; the inoperable status of the radio did not have any effect on the situation.
4. The fire was not caused by arson, vandalism, or any overt human action.
5. The most probable cause of the fire was faulty electrical wiring in the vicinity of the rear passenger door, which allowed an electrical short circuit, with arcing, to ignite the insulation.
6. The fire followed the wiring path, burned through the wall paneling, and ignited a seat cushion; the burning wire also ignited and burned through the roof paneling.
7. Following ignition of the seat, the fire intensity grew rapidly. Large amounts of smoke were generated and fire could not be put out.
8. The nearby store building exterior was heated by the flames and re-radiated heat to the bus in a form of "canyon effect." Only the fortuitous nearness of the fire station and early arrival of the fire trucks prevented a serious building fire.
9. There were no personnel injuries or fatalities.

Scenario analysis indicates that in addition to the wiring that burned, many other wires were run within the bus wall paneling but not protected within metal conduits. Thus, an electrical short anywhere in the wiring could have produced a similar fire, and protection of the wiring is clearly indicated.

Polyurethane seat cushions provided the major heat load and, most important, were easily ignited. As a minimum, the seat covers should have provided resistance to ignition; even better would be replacement of the polyurethane cushions with neoprene foam or seats of more fire retardant construction.

8.1.4.2 Prefire Considerations

Data including regulations, plans and specifications, manufacturers' records and procedures, inspection records and operating records relative to the bus should be gathered. Attention should be directed towards the bases for materials selection, how and where materials were to be used, how the materials were installed, and how the materials were maintained in operation.

8.1.4.2.1 Ignition Source

A listing of potential ignition sources and their characteristics is needed. Such characteristics should include maximum temperature, energy release rate, time of application to target, area in contact, and details of mode of heat transfer.

8.1.4.2.2 Ignited Materials

The target material characteristics are crucial; they must be described in their as-built condition and as they are modified during the ignition and combustion processes. Energy balance at the surface, thermal properties, thickness, and configuration are important.

8.1.4.2.3 Flaming or Smoldering Combustion

Smoldering is characterized by lower spread rates than flaming combustion and, where it exists, usually produces different species of smoke and toxic gases than does flaming combustion. Flaming combustion after smoldering may produce a very rapidly growing fire; smoldering may conceal the ongoing pyrolysis or it may produce smoke that is easily detectable. Thus, the possibility of a material producing smoldering or flaming combustion is an important consideration.

8.1.4.2.4 Fire Spread

The rate of fire spread is essential because it defines the time after ignition when the fire reaches a dangerous size. Spread may occur across the initially ignited material or by jumping a gap to a nearby combustible. It is affected by arrangement, ventilation and flash point of the initially ignited material. (For a secondary combustible it is affected by radiation, gas mixtures and relative position.

8.1.4.2.5 Evolution of Smoke and Toxic Gases

Smoke and toxic gases may provide a means of early detection of the fire, but they also have serious adverse physiological and psychological effects on humans. The evolving characteristics of an ongoing fire may be crucial to human survival or injury and therefore are vital elements in the analysis. Toxic effects and obscuration of vision are particularly pertinent.

8.1.4.2.6 Automatic Detection

So long as smoking is permitted in any part of a bus, automatic detection devices in the passenger compartment will not be effective. When buses were not too large (i.e. one compartment holding less than 40 passengers), detection of fires by the operator or passengers was reasonably effective, but with larger and compartmented buses, some automatic detectors are necessary (and smoking should be prohibited in these areas). It is important that the characteristics of detection devices be determined (if any are fitted).

8.1.4.2.7 Extinguishment

In most buses, the engine compartment has (or should have) automatic or remotely operated extinguishment devices; in the passenger compartment, extinguishment generally is manual. The firefighting agent may be water, CO_2, Halon, or other compounds. The burning characteristics of the polymers involved and the size of the fire will largely determine the effectiveness of the firefighting effort. Fires in the passenger compartments of most buses, heavily loaded with polyurethane material (seat cushions) as well as polymeric side and roof panels, will be extremely difficult to extinguish unless caught early (i.e., in 1 to 5 minutes).

8.1.4.2.8 Flashover

Flashover (discussed in more detail in Section 3.3.9) is a critical turning point in bus fires since it is probable that its occurrence signals complete destruction of the bus, loss of life or severe injury to remaining passengers, and an increased likelihood of the fire spreading to nearby structures. Bus configurations, polymer loadings and bus operating conditions (good ventilation, etc.) importantly bear on flashover, especially where there is low probability of adequate nearby fire suppression. Therefore, bus design changes must be made as required to protect the public and reduce unnecessary losses.

8.1.4.2.9 Spread to Structures and Other Vehicles

A downtown bus fire (e.g., one that occurred in St. Louis, Missouri, May 26, 1976) presents a serious potential for involvement of other structures. In this case, a nearby department store was damaged on the first and second floor. Except for the fortunately early arrival of the fire department, the building could have been heavily involved; the re-radiation between bus and building and structures across the street could well have developed a major fire. It is easy to develop a scenario involving jammed downtown streets, a bus fire in a "building canyon," and fire trucks unable to get to the scene. Five minutes after the bus fire starts, the entire area could be ablaze.

A more likely case of fire spread involves buses in overnight storage that are lined up in close array. A fire starting in one bus could easily progress downwind to all those leeward; in a calm situation, all buses in a row could become involved. The direction and force of the wind, the time of warning and early arrival of adequate firefighting capability determines the degree of involvement of vehicles other than that one in which the fire started. With buses costing from $60,000 to $150,000 and smoke detectors priced at $25.00 to $50.00 per unit, it appears cost-effective to establish adequate fire warning networks in bus terminals.

8.1.4.3 Fire Load

Interiors of buses contain many polymers, all of which contribute to the fire load (see Section 8.2 for materials used). None are as extensive, or contribute so heavily to fire susceptibility as do urethane foam seat cushions (in general use in bus construction). This polymer should be replaced because it is easy to ignite, has a high heat release rate, contributes heavy smoke, toxic gases, and is present in large amounts and general distribution (throughout the interior).

8.1.4.4 Personnel Danger

Under normal conditions, passenger egress from buses through front and rear doors is reasonably good for the physically able. In a rapidly developing, smoky fire, egress through doors and "break-out" windows is more difficult, particularly for the aged and handicapped. Bus design can provide good escape routes (when all functional requirements are considered), and improvement in passenger safety should take the form of increasing the time for egress by reducing the probability of ignition and slowing the spread of fire in the bus.

As in the case of a subway train, a bus fire in a tunnel poses potential smoke and toxicity hazards to other persons in the vicinity. Similar but less severe potentials exist in bus fires in "downtown canyons." Firefighters, however, are at risk wherever the fire occurs.

8.2 Materials Used in Buses

The use of polymers in buses for decoration, comfort, noise reduction and, increasingly, to replace functional metal parts is growing rapidly. Proper selection of materials requires full consideration not only of the functional requirement of the part, but also of the fire performance in every real environment to which the part may be subjected.

Polymeric materials in use in transport systems are discussed in some detail in Chapter 4. Since polymers are being extensively utilized in all new buses, it is helpful to discuss the usages by bus area.

8.2.1 Passenger Compartment Furnishings

8.2.1.1 Seat Cushions and Fabrics

Polyurethane foams are the major materials used in bus seat cushions. They are readily ignitable with small ignition sources and, under certain conditions, even with a cigarette or paper match. Even when covered with a fire-retardant fabric, satisfactory fire performance may not be obtained because of the gradual loss of fire retardancy of the fabric with use and cleaning operations, and the ease of penetration through the fabric (as by a vandal). In many cases, seat cushions are not protected from below, making a small rubbish fire particularly dangerous. A relatively small percentage of buses use fire-retarded neoprene foam seat cushions that evolve dense black smoke and hydrogen chloride gas when ignited but are far less easy to ignite and, when ignited, burn far more slowly than polyurethanes.

Tests by the National Bureau of Standards (1975) on a bus for a metropolitan transit system revealed that:

1. After ignition of a polyurethane bus cushion, smoke generation reduced visibility to zero in 1 to 2 minutes.
2. Polyurethane foam seat cushions involved the rest of the bus in fire much more rapidly than did neoprene foam cushions.
3. Fire growth and spread in the bus is primarily through involvement of the seat cushioning, spreading from seat to seat with little involvement of other interior materials.

A few older buses and some school buses have "hard" seats made of formed (or pressed) fiberglass reinforced composite polymer. These seats are much harder to ignite, burn more slowly and, have a similar heat load. Replacement of these seats with urethane foam cushions in newer buses or in refurbishment produces a very substantial increase in fire risk while producing a small and possibly unneeded improvement in passenger comfort.

Seat cushions are by far the largest heat source in the passenger compartment of modern buses; in most cases, they also are the most ignitable. Since bus seats are subject to vandalism (slashing, etc.), the neoprene interline system does not appear to offer significantly improved fire performance as contrasted to neoprene foam cushioning associated with a fire retardant surface.

In the immediate future, it is suggested that fire-retarded neoprene foam (in lieu of polyurethane foam) be required in seat cushions of new and refurbished buses, that fire-retardant covers be required on the seat cushions, and that a fire barrier be installed under the cushions (e.g., sheet metal, plywood, or a fire-retardant cover) as prerequisite to UMTA funding. Figure 4 demonstrates clearly the improved fire performance of neoprene. In addition, seat construction materials and methods should be developed to provide seats that will reduce ignition susceptibility and fire growth and spread. Reduction of smoke and toxic gas generation should be included as goals of this research and development effort. This research and development effort probably

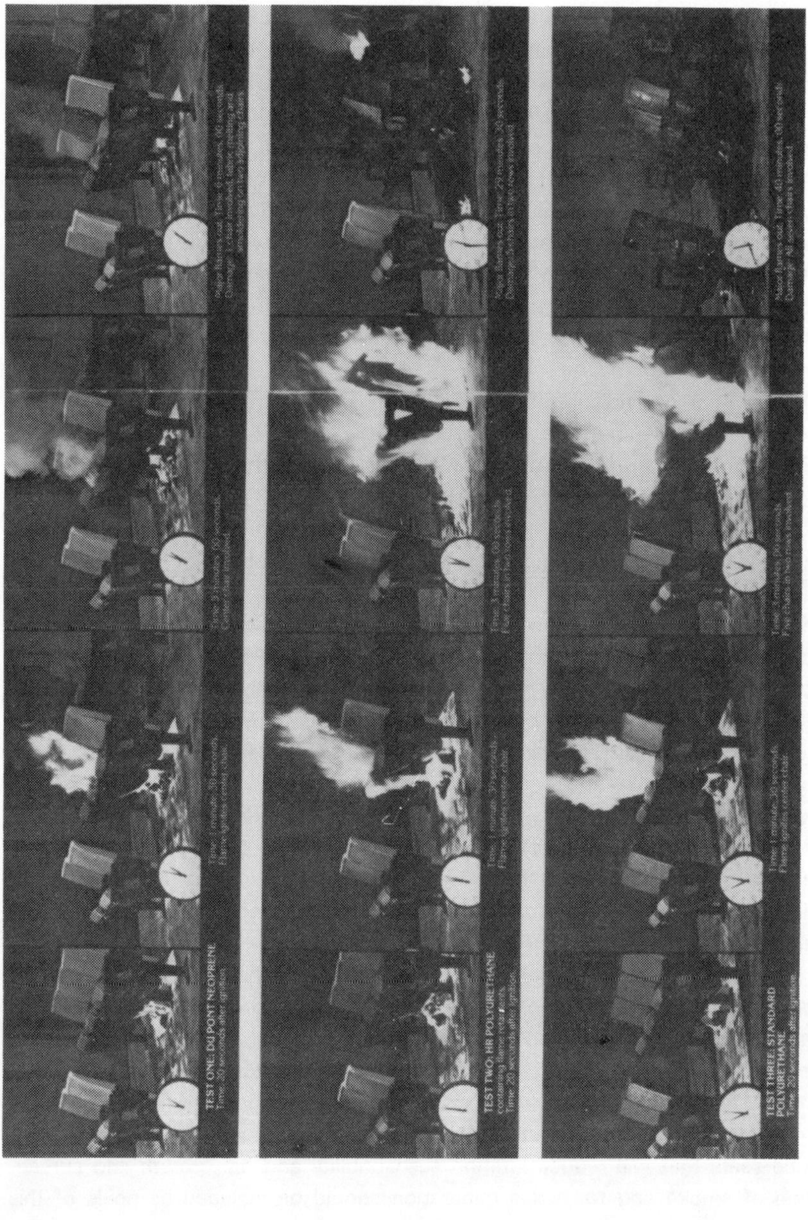
TEST ONE: DU PONT NEOPRENE
Time: 20 seconds after ignition
TEST TWO: HR POLYURETHANE
containing flame retardants
Time: 20 seconds after ignition
TEST THREE: STANDARD
POLYURETHANE
Time: 20 seconds after ignition

Figure 4.

should be primarily funded by DoT, since the results will be applicable to all forms of mass transportation.

A number of fire-retardant fabrics and coverings are in use. Most resist small ignition sources (matches and cigarettes), but are susceptible to ignition from small trash fires. Some fabrics lose their fire retardancy characteristics during laundering and cleaning. PVC-coated fabrics are durable and reasonably fire resistant, but release high density smoke and hydrogen chloride gas when burning. Nomex® has desirable fire performance, but costs have held back extensive use. The development of more suitable fabrics is desirable.

8.2.1.2 Seat Frames and Shrouds

Seat frames usually are metal but often are made of fiberglass-reinforced plastic. Since portions are exposed, they should meet the same flammability and smoke standards as the wall and ceiling panels. Backs and sides of seats are covered with material (the shroud), and PVC mixtures and coatings often are used in this connection.

8.2.1.3 Carpeting, Floor Tiles, and Mats

Carpeting, floor tiles, and mats are specified by various operators for floor coverings. Neither vinyl composition tile nor rubber tile nor matting has been shown to be a serious fire hazard. Carpeting (nylons, acrylics, wool, etc.), installed principally by decorators to add "plush," have a wide range of flammability. The NBS Flooring Radiant Panel Test can be used to determine the best carpet materials from a fire retardant standpoint. Considering initial cost, wear and soiling, there appears to be little reason to specify carpeting rather than the far more fire-resistant tile or rubber matting floor coverings that have been used for a number of years.

8.2.1.4 Wall and Ceiling Panels

Metal wall and ceiling panels have gradually been replaced by formed polymer composites based on PVC-acrylic, PVC-ABS and melamine resins. These composites are not easy to ignite but will burn with substantial heat, smoke, and toxic gas release in a fire environment (e.g., that of a burning seat). A sustained ignition source (e.g., an arcing electrical short) also will ignite these panels.

Decorators have applied fabrics and carpeting to wall polymer composites. Carpeting, and most other fabrics, are much easier to ignite and will spread fire much more rapidly in a vertical orientation (than in horizontal), thus adding appreciably to the fire risk. It appears that the decorative effect (and some noise attenuation) is not worth the added fire risk.

8.2.1.5 Lighting Diffusers

Lighting in buses generally is provided by fluorescent lamps screened by plastic diffusers that often are made of polymethylmethacrylate. Although flammable, they usually are not involved in the eary stages of fire (ignition, fire spread) but melt and drip as the heated gases rise, thus adding to the fire load.

8.2.1.6 Advertising Panels and Matter

Panels used for advertising may be transparent and cover a paper advertisement or have the advertisements printed directly on them. They are often of styrene/butadiene composition coated with a co-polymer. The total heat load involved is small, and their contribution to the early stages of a fire is limited when the advertising panels are located well above seat levels. Panels at seat level (and at car ends, doors, etc.) could provide good paths for fire spread or for fire initiation where such panels cover exposed electrical wiring. In the lower level advertising panels, it appears that clear polycarbonates would provide improved fire characteristics at reasonable costs.

8.2.1.7 Padding on Metal Frames and Stanchions

Metal frames, stanchions, and sharp corners often are padded for the protection of passengers. Urethane foams are used in many cases. The comments regarding unsatisfactory fire performance, when used as seat cushions, pertain. Safer foams, such as fire-retarded neoprene foam, or other improved paddings should be used.

8.2.2 Passenger Compartment Services

The supply of necessary services to the passenger compartment provides sources of fire ignition, spread, and support. Such services include heating, cooling, ventilation, lighting, solace (as in public address and music systems), and equipment operation. Thus, proper materials are required in these service-providing systems from a functional point of view, and proper operational design and procedures to reduce fire hazards are vital.

8.2.2.1 Air-Conditioning Ducts

Air-conditioning ducts in the passenger compartment also are used for heating and ventilation. Materials must be stable and fire resistant under cooling and heating conditions and should not spread fire fore and aft in the bus by virtue of the flammability characteristics of the ducting material (i.e., the fire resistance of ducting material must be equal to or better than that of the wall and ceiling panels, and the adjacent insulation). In particular, when passing through fire bulkheads, the ducting material must not transmit the fire from one side to the other by burning. Further, it also would be desirable that the ducts have closures that could be closed at the onset of a fire by moving a damper manually or, better, automatically when the circulation fans were shut off as part of the bus operator's standard fire procedure.

8.2.2.2 Electrical Wiring

Wiring is used to support many functions such as equipment operation (fans, closures, etc.), lighting, music and public address audio, and operating instruments. It carries different amounts of current throughout the bus, often in exposed wireways and sometimes not in harnesses. Although some designers specify conduit enclosures for power wiring, this is not always the case. Chafed or overloaded wires can develop short circuits that become ignition sources often in proximity to polymer wall and ceiling panels or thermal insulation. (It is noted that one school bus manufacturer is not enclosing all power and lighting wiring in conduit or equivalent.)

Wiring insulation burns and may itself be a source of fire transmission. Currently, insulation materials in use are crosslinked polyolefins, PVC, etc. They are not easy to ignite from the outside, but, given a sufficient period of ignition time, they will ignite and will readily transmit fire to other polymers (including other wires in the same harness). While proper fusing of power and lighting circuits normally takes care of short circuits some overloads, vehicle fire statistics show many electrically initiated fires. Thus, it appears appropriate to suggest conduit for lighting and power circuits and usage of superior wire insulations developed for aircraft where conduit appears infeasible.

8.2.2.3 Music and Public Address System

This system, where installed, usually has little effect on fire hazard. The speakers, made of plastics, are located in elevated positions, usually away from most potential ignition sources; the wiring carries small currents unlikely to produce ignition even under short circuit conditions.

8.2.2.4 Door and Other Mechanism Operators

Door operating mechanisms may be energized electrically, hydraulically, or pneumatically. The mechanisms themselves are seldom ignition sources but can cause problems in the energy supplying system. Electrical door actuators can overload the wiring circuits with consequences addressed in Section 8.2.2.2. Hydraulic systems may "diesel" or explode, sending mists of hot hydraulic oil onto light bulbs, etc., with consequent fires. Fortunately, the number of hydraulic systems inside bus passenger compartments is diminishing. Where still in place, specification of fluids per ANSI B93.5, Practice for the Use of Fire-Resistant Fluids for Fluid Power Systems in lieu of petroleum based hydraulic fluids, will markedly alleviate the hazard.

Air systems seldom initiate or support the spread of fires. Both air and hydraulic systems use gaskets, lubricants, etc., between metal piping and other parts, but these are small in size, number and well separated. Virtually any gasket material which is functional is adequate in fire performance.

8.2.2.5 Bus Operator Controls and Devices

The bus operator's station contains large numbers of knobs, gauges, dials, instruments, controls, and other devices necessary to run the bus. Many wires, pipes and levers made of or containing polymers are used. Fortunately, the available energy inputs are low and the materials are not highly susceptible to ignition and are generally under direct observation. Few, if any, fires start here; these are readily put out.

8.2.3 Bus Body, Including Passenger Compartment Structure

Until recently, bus bodies were primarily of metal construction. Increasingly, parts and sections are being made of polymers with consequent weight and cost reduction and many other benefits. It is expected that, within a few years, a large portion of the body and its accoutrements will be polymeric in nature (e.g., the bumpers, bumper beams, radiator and fender supports, hood liner panels, door impact beams, doors, transmission support). Soon thereafter will follow the chassis, drive shaft, differential, and rear bulkheads. If these substitutions are made with proper consideration for fire performance, a substantial advance will be achieved (for a good overview, see Brake 1976).

The discussions that follow relate to current bus construction.

8.2.3.1 Chassis

Chassis currently are made of metal forgings, weldments, or combinations thereof. Fiber-reinforced composites now are being substituted on an experimental basis, particularly at the fore and aft extremities. Within a few years, it may be possible that the entire chassis will be polymeric in nature. Obviously, immediate concern must be given to the chassis parts in the vicinity of the hot exhaust system to ensure that the chassis will not catch fire. Similarly, hazard studies and risk-cost trade-off studies will be needed to evaluate the benefits of structural integrity in fires as provided by metal chassis versus the reduced costs, weight, and other advantages of polymers.

8.2.3.2 Floor

In most bus designs, the floors not only support the passengers but also perform functions relating to longitudinal and transverse structural strength, provide separation of certain auxiliary systems for the passenger compartments, provide reduction of noise from machinery and air flow, and serve as fire barriers. Thus, not only is it important that the floors be strong enough, but also that they have minimum and properly designed penetrations and that they be fire resistant.

Floors in the past were made of steel plating. Later, a sandwich of galvanized steel sheet, plywood, and steel sheet was substituted; this construction is being extensively used today. Both constructions were and are satisfactory.

Certain manufacturers replaced the steel-plywood-steel sandwich with a steel-polyethylene-steel sandwich; the fire performance has been shown to be inferior to the steel-plywood-steel combination. In addition, other manufacturers make floors of plywood without steel covers, which also appears to be a questionable practice.

Since significant gains may be made by use of synthetic polymers in lieu of current construction, increased efforts in research, experimentation, and development are warranted; however, none of the experimental systems should become part of actual bus floor construction methodology until adequate tests are developed and specified and the materials qualified by those tests.

Auxiliary equipment often is installed below floors (e.g., exhaust system components, batteries, air compressors and air lines, hydraulic pumps and liners). In trolley buses, electrical connection boxes and switching and power devices are so installed. All these items provide substantial heat in normal operation and are potential ignition sources of large magnitude during abnormal operations. The floor materials and construction must provide adequate resistance to the substantial operating hazards.

Design of floor penetrations is important to the maintenance of floor integrity as a barrier to heat, fire, and fire products. Improperly used polymeric penetrations in assemblies can contribute to serious degradation of performance of an otherwise satisfactory floor.

Until acceptable definitive standards are developed and specified, polymer compositions should not be used unless the fire characteristics are shown to be equivalent to those of the metal components being replaced.

8.2.3.3 Wheel Well Covers

In most mass transit buses and school buses, the wheel well covers become a portion (curved) of the floor itself; interstate buses may have a continuous metal floor above the upper tangent of the wheel well covers, in which case the covers become an end bulkhead for baggage compartment, engine space, etc. In some cases, the covers on mass transit buses have been made of fiber-reinforced polymers formed into sheets, which has reduced cost, weight, and corrosion.

Under normal conditions this type of design functions well, but when the tires or brakes catch fire, polymer wheel well covers are subjected to a large continuing ignition source and will catch fire, burn through into the passenger compartment and set the seats on fire, resulting in complete destruction of the bus (see, for example, National Transportation Safety Board, Recommendation H76-7, 1976). Considering the relatively small value of the benefits obtained and the substantial reduction of fire performance (and consequent increase in losses), it is believed prudent that polymeric wheel well covers should not be used in newly constructed buses unless the fire characteristics are shown to be equivalent to metal covers.

8.2.3.4 Side Structure

Fiber-reinforced composite polymers are increasingly replacing metal units in side panels, impact beams, and stiffeners as well as windows and door supports. Comments made for floors (Section 8.2.3.3) are generally applicable here, except that the outer skin of the bus side operates in a less challenging environment than the underside of the floor. There are no nearby heat sources and no large ignition sources. However, the side structure contributes to the strength of the body; it should retain its strength and not catch fire from a localized ignition source (e.g., a pile of trash inside or outside the bus). Since adequate tests have not been developed or specified, it appears that the outer shell structure should be made of metal until polymer composites of equal fire performance can be developed.

Experimental buses using fiber-reinforced polymer side structures have been built. At least one (UMTA prototype, 1975) has burned completely from fire spread along the side structure even when the ignition source (engine compartment) was thought to be protected (from the passenger compartment) by a metal fire wall. This fire not only indicated the poor fire performance of the fiber-reinforced polymer selected, but also indicated the need to extend the fire wall into the side structure for fire protection effectiveness. The Wayne Company's "Lifeguard School Bus," is an example of a bus that offers many advantages in body construction and fire performance (except where urethane foam seats are fitted) including full-length steel exterior body and roof panels, full-length interior panels, steel body frames, electric wiring in conduit (in passenger compartment), and seat bottom covers.

8.2.3.5 Roof Structure

Roof structural panels are increasingly being made of fiber-reinforced polymers. These provide a less costly, low weight, high performance solution of great attractiveness. Until the seat cushion fire hazard situation is greatly improved (see Section 8.2.1.1), this situation may be satisfactory for internal combustion engine buses. When the fire hazards of the seats are corrected, it will be desirable to require the same fire performance standards for roof structures as for bus side structures.

The trolley bus is a special case where the trolley and the electrical wire it contacts may malfunction to produce a high energy, continuous ignition source on the roof of the car. In this case, the same reasoning applies as for the floor of the buses; only a metal or metal-clad composite roof with a thin insulation material covering added on top of the metal is acceptable.

8.2.3.6 End-Caps (Front and Rear) and Other Shell Structures

The forward end-cap for buses with engines in front is not only an end structural closure but also a fire wall separating the engine from the passenger compartment. The same basic situation as for the floor exists; this end-cap should be metal (for the present). The rear end-cap should comply with the same fire standards as the bus sides. The forward end-cap for buses with engines in the rear should follow the standards for bus sides; the rear end-cap, a fire wall with specially designed passthroughs, should adhere to floor standards. Other shell structures (covers, baffles, partitions, etc.) should follow floor or side fire standards, depending on their proximity to ignition sources and their proclivities to support the spread of fire. Trolley bus end-caps should be in accordance with fire standards for the sides.

8.2.3.7 Insulation, Thermal and Acoustic

A single material usually is used to perform both as thermal and acoustical insulation, filling the space between inner and outer walls and ceiling and between floor frames.

Sprayed-on polyurethane foam has been used extensively. It is easy to ignite and, while burning, contributes significant amounts of heat, smoke, and toxic gases. Where the wall and roof spaces are part of the ventilation system plenum, undesirable fire products can be rapidly distributed throughout the bus.

Fiberglass batts also have been used. Where not more than 2 to 3 percent resin binder is used, satisfactory fire performance has been obtained.

8.2.3.8 Windows, Glazing, Seals, and Frames

Plastic windows have replaced glass where vandalism has been a problem; both polycarbonate and polymethyl methacrylate windows have been used. In some cases a single plastic sheet is employed; in others, double glazing with the plastic outside, an air gap, and safety glass inside is used.

Polycarbonate glazing is superior to polymethyl methacrylate. It appears well worth the additional cost involved (see Section 7.2.5 for a discussion of the subject).

Window seals and gaskets currently are less resistant to fire than the windows and side structure. Improved materials will be needed if the seat problem is corrected.

Frames for windows, usually metal in the past, are increasingly made of formed, reinforced polymers. The fire standards should be similar to those for the side structures.

8.2.3.9 Doors

Doors consist of frames and glazing and should comply with side structure fire standards.

8.2.3.10 General Comments on Body Components

There are a number of conflicting functional demands on the body structure which require trade-off compromises. In the above discussion no attempt was made to trade-off the functional requirements against the fire performance needs. It appears that the real fire performance requirements have not been adequately considered in many substitutions of polymers for metals; it appears also that the fire characteristics of polymer composites are inadequately known to designers, builders, and operators of buses.

8.2.4 Engine Compartment

The overwhelming fire consideration in the engine compartment is the prevention of fuel leaks or explosions (fuel is excluded from the committee's deliberations). Although there are many engine compartment fires from failures of various components, a well-designed, well-maintained engine compartment will provide fire safe operation using currently available materials. Further, some engine compartments are provided with remote or automatically operated fire extinguishment devices that generally work well in the relatively small spaces involved. There have been few personal injuries or fatalities from engine compartment fires except where the fires have been spread through poor body design or passenger compartment furnishings and services designs.

8.2.4.1 Engine and Auxiliary Power Devices

Engines and auxiliary power devices are almost entirely metal and probably will remain so for the foreseeable future. They are believed to have satisfactory fire performance.

8.2.4.2 Gaskets, Fuel Lines, Hoses, Belts, Etc.

Functional performance of these items is paramount. There is little or no evidence that they have initiated fires, and they have been major contributors to the spreading of fire only when they ruptured and allowed fuels to leak out.

8.2.4.3 Fuel, Lubricating Oil, Hydraulic Oil

These items are excluded from committee consideration. It is noted that the more expensive fluids per ANSI B93.5 Practice for the Use of Fire Resistant Fluids for Fluid Power Systems do provide a substantial increase in fire safety.

8.2.4.4 Structures and Shields

Most baffles, supports, etc., should have the same basic standards as the bus side structure. Those items supporting batteries, electrical power (or switching) devices, and exhaust system components should have high fire performance characteristics at least equal to those of the bus floor structure.

8.2.5 Exterior of the Bus

A number of systems and components are attached to the bus structure but are outside of it. Most are currently of metallic construction but, being heavy, are early candidates for replacement by formed reinforced polymers.

8.2.5.1 Exhaust System

Exhaust system are all metal. Considering their temperatures in normal operation, it appears unlikely that these components will be early candidates for polymer substitution. The heat thrown off by the exhaust system is, however, a major factor affecting material selection for other exterior parts (this heat flux is a very important consideration for rear engine compartment buses where the exhaust system is inside the engine compartment).

8.2.5.2 Power Transmission

Mechanical and hydraulic transmissions, power shafts, differentials, axles, and axle housings are largely metal but are early candidates for replacement by polymers. Proximity to heat sources in normal operation or heat resulting from malfunction (brake overheating, tire overheating, etc.) should determine the fire performance standards required.

8.2.5.3 Brake Systems

Except for the polymeric wearing shoe or pad, braking system components are metallic. Overheating of the metal structure and consequent combustion of the brake shoe (pad) can involve tires, the hydraulic fluid (in the brake actuator), and the bus floor structure. This potential heavy ignition source must be considered in design (materials selection, form, cooling, etc.) of wheels, axles, and floors.

8.2.5.4 Wheels and Tires

Wheels, now all metal, are early candidates for polymeric substitution. Considering the heat and ignition sources from brakes and from underinflated tires (tire fires), polymers used in wheels should have extremely high fire performance (not so much as to prevent the spread of flame but to ensure that the wheel will not collapse while under heat [flame] attack with consequent loss of bus integrity).

Trolley buses and other vehicles having electric power motors at the wheel have another potentially large and continuous ignition and fire source; the above discussion applies. Tires are not problems unless improperly operated (underinflated, rubbing on other tires or structures, etc.) or ignited from other sources.

8.2.5.5 Fuel Tanks and Fuel Piping Systems

Some metal fuel tanks have already been replaced by polymer composites (nylon, polyethylene) and have operated satisfactorily where design (materials selection, location, etc.) has been good. The materials used, however, are susceptible to heat degradation (e.g., from a brake fire, exhaust system, or other unexpected fire) leading to the possibility of loss of tank integrity, fuel leak, and catastrophic consequences. If unexpected fire is severe enough to cause explosion of the fuel, no tank of any construction will suffice. Fuel piping systems generally are made of metal except in some portions of the engine compartment. Little incentive exists to shift to polymeric tubing.

8.2.5.6 Suspension Systems

While most suspension systems are made primarily with metal springs, shock absorbers, etc., some buses have pneumatic suspension systems utilizing air bags and other components. To date, these systems have not been observed to be sources of fire initiation or fire spread.

8.2.6 Improved Bus Construction

Some bus operating authorities have begun to specify the UMTA "Guidelines" for materials used in bus construction. This practice has resulted in considerable improvement from a fire safety standpoint.

The fire scenario in Section 8.1.5 developed for a bus built to MVSS 302 and using urethane seats resulted in the total loss of the bus. A fire scenario on an improved bus, built to UMTA guidelines, resulted in only a minor fire spread despite a large ignition source. This latter scenario is described in detail below.

The vehicle involved was a 1976 Flxble bus that had been in revenue service for about 4 months. All interior materials conformed to the UMTA "Guidelines for Flammability and Smoke Emission" (with exception of the fabric-backed vinyl upholstery which exceeded the smoke emission standard).

On the night of the fire, the bus was being returned to a garage by a mechanic after the operator reported transmission trouble. The bus was operating normally when the mechanic observed through his rear view mirror smoke coming from the engine compartment.

The fire did severe damage to the engine compartment. Most of the wiring insulation was burned away as were most other nonmetallic parts. In addition, a metal rear control box located in the right rear of the engine compartment was half melted. Damage to the bus interior was confined to a wire duct in the right rear corner plus the burning and melting of the insulation and vapor barrier under the rear seat. There was no flame spread within the interior or involving interior materials.

When the fire department arrived on the scene, they reported that the engine compartment was flaming intensely even though the mechanic had attempted to extinguish the fire with the on-board dry chemical fire extinguisher. There are no accurate time estimates between the discovery of the fire and its extinguishment by the fire department. (An opinion would be that the time could not have been very long due to rapid reporting and nearness of the fire station.)

While the exact cause of the fire was impossible to determine, it was the consensus of those involved that the cause was electrical. Since the transmission malfunction also was electrical, it was assumed that this fault may have been involved in the fire cause.

The significance of this fire is that a similar fire in December 1974 involving another manufacturer's bus built to MVSS 302 resulted in severe damage to the bus interior and burning through of the aluminum roof. This fire also was discovered quickly with quick reponse by the fire department. The cause was electrical and in the engine compartment. It was suspected that the first materials ignited were the rear seat and polyurethane air-conditioner pad. These materials resulted in rapid flame spread and rapid heat release versus the absence of flame spread in the Flxble bus with its neoprene cushions and other improved interior materials.

8.2.7 Summary

Polymer systems have been introduced as replacements for metal in bus construction, generally in full compliance with MVSS 302. MVSS 302 is inadequate to provide bus passenger safety in fire situations. Polymer systems have been introduced without adequate consideration of the conditions that should be expected in fire situations. Fire scenario development and analysis, not currently required, is suggested as a forward step in this regard. Polymer systems have been introduced without adequate knowledge of their fire behavior, particularly regarding generation of smoke and toxic gases and the synergistic interactions of other existing polymers in use. Polymer systems provide significant public benefit in bus systems, and new polymer systems should be encouraged, developed, and installed when fire safeguards are provided.

8.3 Tests

The standard most prescribed for bus construction is MVSS 302, which in essence requires that a material used in a bus have a flame propagation rate of less than 4 inches per minute while in the horizontal position. This simplistic standard may screen out a few obviously unsatisfactory materials, but it is inadequate, by itself, to provide a degree of fire safety commensurate with the public need in buses.

Certain operating authorities, (Boston MBTA, New York, etc.) specify other tests with some resulting improvement in fire safety.

The National Transportation Safety Board has discussed MVSS 302 in a number of its reports including the following comments from Safety Recommendations H-75-12 and H-75-13 (1975):

> . . .The Safety Board is concerned not only with whether the interior materials comply with FMVSS 302, but whether the standard is adequate to provide protection for passengers. The Board's primary concern is the time required to evacuate the vehicle before an environment which will not support life is produced within the vehicle . . .
>
> . . .FMVSS 302 fails to provide predictable fire propagation information. Therefore, it does not provide safety information usable in predictable, fuel fed fires which follow motor vehicle accidents . . .
>
> . . .Though few in number, accidents involving fire are not only much more severe and costly than the average bus accident, but the number of fatalities is 30 times the number per accident for all bus accidents . . .

Recommendations H-75-12 and H-75-13 suggest testing materials in attitudes that could be expected in accidents and expansion of MVSS 302 to include vertical burn tests of all vehicle interior materials (similar to FAA requirements in 14 CFR 25.853).

The generally unsatisfactory situation with regard to test methods, specifications, and standards is discussed in Chapter 5 and, more fully, in Volume 2.

8.4 Smoke and Toxicity

Smoke and toxicity are discussed in Chapter 6. Testing for smoke and toxicity is discussed in Sections 5.4.6 and 5.4.7 respectively of this volume and detailed in Volume 3 of the committee's report. It appears safe to say that smoke and toxicity generation from combustion of polymers is receiving greater attention as the adverse effects are being more widely recognized. In the case of toxicity, an adequate standard is not yet available.

8.5 Related Considerations

8.5.1 Statistical Base

The statistics available relative to bus fires are not complete and are not drawn from the same base or the same rules. Fire experts are too seldom involved in fire reporting. At best, the statistics are a guide as to the nature of the problems rather than to the magnitude of hazards. Qualitative conclusions are the best that can be drawn. Pending development of a reasonable statistical base, scenario analysis appears to be the best approach to improvement of fire safety in buses.

8.5.2 Risk Analysis

Fire risk analysis (hard analysis) is, so far as the committee could determine, not a part of the specification or design procedure for buses. When performed, it is more often intuitive and subjective than scientific and objective. In an industry where low bidder wins and where there are no specification requirements for risk analysis and no risk-reward ratio methodologies, the manufacturer performing such analyses and modifying his design to incorporate the results would not submit the lowest bids and, ultimately, would go out of business. When bus operating agencies, owners, and regulating agencies develop requirements and specifications for fire safe buses, manufacturers will build them.

8.5.3 Use of Materials Without Tests

New polymers are being synthesized in laboratories and developed into manufacturing materials with adequate testing relative to *functional* use (tensile strength, fatigue, creep, etc.). If these materials pass MVSS 302, already described as inadequate, they may be, and are being, introduced into bus construction without adequate testing (and therefore knowledge) of their *fire behavior*, including smoke and toxic gas generation. Knowledge of or adequate prediction of fire behavior appears to be a needed mandatory addition to engineering handbooks before a material is used.

8.6 Conclusions and Recommendations

8.6.1 Matters of High Priority Attention

Conclusion: Polyurethane foam seat cushions are a serious fire hazard in contemporary buses.

Conclusion: Floors and other structures serving as fire barriers have some penetrations that are poorly designed or are constructed of unsatisfactory materials (from a fire safety standpoint). The most serious situation in this regard is the use of fiber-reinforced polymer wheel well covers.

Conclusion: Electric wiring, particularly that which carries current to operating devices (and thus is susceptible to large short circuit current). concealed behind interior body paneling is a potentially grave and strong ignition source. Its proximity to polymeric insulations and interior paneling adds to the hazard.

Recommendation: Prohibit installation of polyurethane foam seat cushions in new buses, and refurbish current operational buses with better materials. Specify a better material, such as fire-retarded neoprene foam, until such time as a more suitable material than neoprene foam becomes available.

Recommendation: Withhold grants to bus operating authorities for new bus purchases (including school buses) until purchase specifications are changed.

Recommendation: Require that wheel well covers made of fiber-reinforced polymers provide fire resistance equivalent to that of steel to prevent wheel (tires and brakes) fires from penetrating the passenger compartment. Specifications for wheel well covers should match those of the structural floor regarding fire performance.

Recommendation: Run electric power wiring, insofar as practical, in metal conduits. When this is not possible, the routing of such wiring should be carefully designed and monitored in the manufacturing process to ensure that the wiring is protected from cuts and chafing and that it is properly fused. It should be protected from contact with polymers used in insulation and body paneling, and the insulation of the wire should have good fire performance characteristics.

8.6.2 Items for Progressive Improvement of Fire Performance of Buses

Conclusion: The use of composite polymers having good fire performance offers many advantages, particularly in bus construction.

Conclusion: The polymers currently used in bus bodies do not have good fire performance.

Conclusion: Experimental and prototype buses (UMTA and manufacturer supported) have not been subjected to rigorous fire performance analysis using scenarios; limited scenario analysis leads to a tentative conclusion that materials selected and used in such experimental buses will not give adequate performance for fire safety.

Conclusion: Many polymers are not being properly used in bus construction (e.g., carpeting used on side or end walls or on the ceiling).

Recommendation: Initiate a program to develop, qualify, and approve polymers having good fire performance for bus construction.

Conclusion: Replace the polymers to be used in bus bodies with steel or fire-retarded composites providing equivalent fire protection.

Conclusion: Require a rigorous fire-biased analysis, using scenario technique, for each new bus design. Withhold funds for development or acquisition until such analysis quantifies the risks and shows that they have been reduced to an acceptable level.

Conclusion: Analyze new and existing bus designs with the aim of eliminating improper or unnecessary use of flammable polymeric materials (e.g., the use of carpeting in vertical or overhead positions).

CHAPTER 9

PASSENGER AUTOMOBILES

9.1 Introduction

9.1.1 Scope

The fire safety hazards, both real and potential, associated with the use of polymeric materials in passenger carrying automobiles are discussed in this chapter.

9.1.2 Nature of the Fire Hazard in Automobiles

Although the incidence of fire in automobile accidents is low (0.5 percent), the number of fires in passenger automobiles has been increasing steadily. Trisko (1975) indicates that from 1968 to 1973, the estimated number of automobile fires increased at an average annual rate greater than 10 percent while structure fires increased at an average annual rate of less than 2 percent. Since 1972, the number of automobile fires has exceeded 500,000 per year. The increases in the number of vehicle fires and in dollar losses are shown in Table 1.

Table 1. Estimated Vehicle Fires and Related Losses

Year	Number of Vehicle Fires	Losses (million dollars)
1971	482,400	96.54
1972	550,300	120.30
1973	574,000	125.00
1974	594,000	135.00

Note: Data from National Fire Protection Association, Boston, Mass., 1975.

Losses have continued to mount despite the promulgation of MVSS 301 in 1968 as indicated in Figure 1 (Lauriente and Wiggins 1976). This standard requires that passenger car fuel systems be designed to withstand a 30 mph frontal collision without experiencing fuel losses exceeding 1 ounce during or after impact. Lauriente and coworkers also concluded from their study that the fire cost ratio has increased 110 percent faster than the gross national product during the past 5 years, indicating an undesirable trend for automobile fire safety in the future. Recent statistics published by the Insurance Institute for Highway Safety (1975) also indicate that motor vehicle fires are increasing 10 percent annually, a rate five times greater than the rate of increase for building fires.

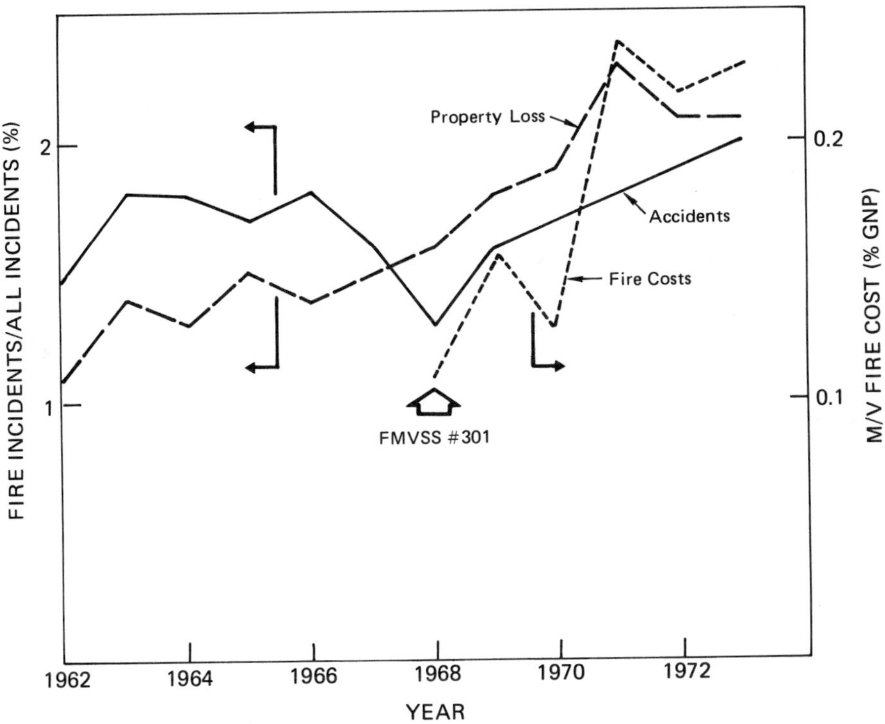

Observations:

- Motor Vehicle Fire Accidents Are Rising
- Property Loss Has More Than Doubled in Ten Years
- Fire Cost Ratio Has Increased 110% Faster Than the GNP During the Last Five Years
- Motor Vehicle Fire Losses Have Increased Despite FMVSS #301

Sources:

- Accident Facts
- Fire Journal
- Statistical Abstract of the United States

Figure 1.

Statistical data for the number of post-crash fire victims have been difficult to determine. Cooley (1974) reported that these difficulties exist because some researchers tend to fail to distinguish the number of fatalities in automobile accidents accompanied by fire (i.e., the rate or frequency of fatalities resulting from fire in motor vehicle accidents) and the number of fatalities resulting from fire alone. Cooley estimated the number of motor vehicle fatalities that would not have occurred, had a fire not started, to be from 450 to 650 in 1972 alone.

Johnson and Sanderson (Trisko 1975a), however, reassessed the data accumulated by Cooley (Figure 2) and estimated that the number of motor vehicle fatalities occurring as a direct result of burns nationwide for 1972 ranged from 625 to 1430. In addition, they estimated that the number of nationwide fire accidents during 1972 ranged from 5,000 to 10,000, but these figures may be considerably higher "since the data base for these studies was drawn largely from reports and statistics which were not specifically designed to report vehicle crash fires."

A recent NFPA report (1973) estimates that more than 450,000 fires occurred in private passenger cars and trucks in 1971, with an average loss of $200. It also was estimated that between 3,500 and 4,000 deaths were caused by those fires. These data indicate the fire hazard to be even greater, but the figures probably do not account for those deaths not directly attributable to fire as indicated by Cooley. From the above analysis, it may be concluded that motor vehicle fire losses account for about 500 fatalities and about $130 million of losses per year.

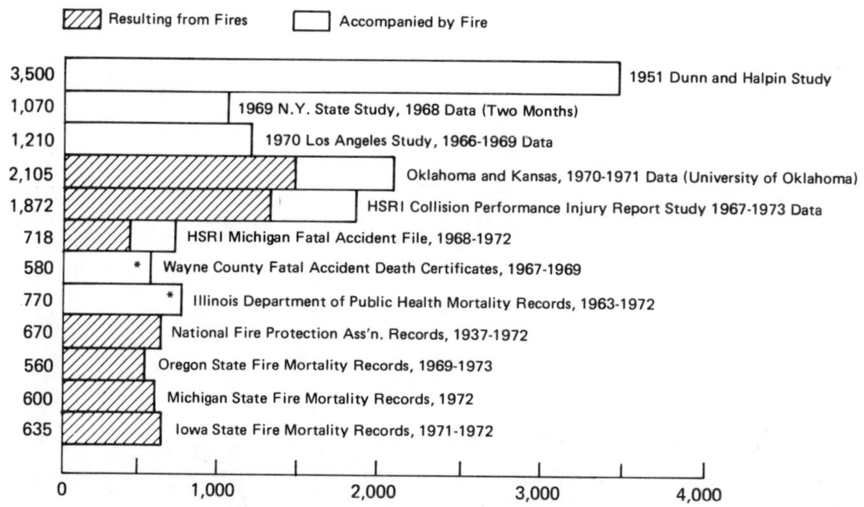

Source: HSRI Special Report, Fire in Motor Vehicle Accidents, April 1974.
*Report text indicates these were fatalities resulting from fires.

Figure 2. Estimated Annual National Fatalities (1972).

Automobile fires, as other fires, require a flammable substance, an ignition source, and oxygen. Almost all of today's automobiles carry a flammable substance — gasoline or diesel oil — and have materials, such as those used for vehicle upholstery, insulation, plastic bodies, trim parts, etc., that can fuel a fire. Ignition can result from many sources such as:

1. Electrical short circuits or other electrical malfunctions that cause excessive heating of conductors of components.
2. Sparks from the engine ignition system.
3. Hot exhaust system components.
4. Backfiring of the engine.
5. Overheating of tires, brakes, and bearings.
6. Sparks generated by friction from a collision or by metal components scraping against the pavement.
7. Smoldering of cigarettes or other smoking materials.

Sufficient oxygen is usually available for the fire. Table 2 provides data from a study of the frequency of passenger automobile fires by incident and area of fire origin.

Table 2. Frequency Distribution of Passenger Car Fires by Type of Incident and Area of Fire Origin, 1973 National Survey of Motor Vehicle Fires

	Type of Incident			
	Noncollision		Collision-related	
Area of Fire Origin	**Number of Fires**	**Percent**	**Number of Fires**	**Percent**
Engine	1,085	59	39	54
Passenger	647	35	3	4
Fuel Tank	59	3	24	33
Trunk	31	2	3	4
Tire/brake	29	2	3	4
Total	1,851	100[1]	72 [1]	100[1]

Note: Data from Trisko 1975b. These figures do not include fires of unidentified origin. The 1,923 passenger car fires in which the area of origin was identified comprised 83 percent of the 2,325 passenger fires surveyed, and 73 percent of the 2,637 total motor vehicle fires surveyed.

[1]Column does not add to total due to rounding.

These data indicate that collision-related fires are largely fuel fed (i.e., fires related to the engine or fuel tank) with more than 85 percent of the total falling into this category. Such fires, of course, are primarily related to the flammable fuel and probably are only secondarily related, if at all, to the type and quantity of polymeric materials. In non-collision-related fires, however, 35 percent of the fires originating in the passenger compartment are primarily related to the materials used to construct the passenger compartment. The flammability characteristics of polymeric materials are a major factor in these fires since these materials now are by far the predominant materials used in interior furnishing (see Section 9.4).

No general statistics are currently available concerning the cost and frequency of non-collision-related passenger compartment fires. A limited number of data have recently become available from the City of Los Angeles Fire Department (1974) and is reproduced in Table 3.

Table 3. Origin of Non-Collision-Related Fire in Passenger Automobiles in Los Angeles, California.

Engine Compartment	3,984
Fuel Compartment	138
Passenger Compartment	1,717

Despite the absence of property damage and injury statistics, these data indicate that fires originating in the passenger compartment comprise about 30 percent of the total reported automobile fires in the city during that year. Although the data are not sufficient to justify any firm conclusions concerning the fire safety rating of automobile interiors, they do indicate that attention to fire-hardening of passenger compartment interiors is worthwhile.

9.2 System Design, Operation, and Fire Safety

9.2.1 Introduction

Although the flammability characteristics of an automobile are primarily related to the flammability of the materials of construction, it has been well established (see Chapter 3) that fire safety or fire hazards can be realistically assessed only by consideration of the total system rather than consideration of the individual materials used to construct the system. This conclusion is easily illustrated by analysis of the data summarized in Table 9 that show that more than 50 percent of all vehicle fires, collision-related or non-collision-related, occurred in the engine compartment of the vehicle. The ssecond most important site of non-collision-related fires (35 percent) was found to be in the passenger compartment while the third largest group of collision-related fires related to the fuel tank. One cannot realistically assess the relative fire hazard associated with fires in the engine compartment unless one considers the entire system (i.e., ignition source, material and fuel flammabilities and the relative proximity of these materials to the ignition source).

9.2.2 System Design Considerations from a Fire Safety Standpoint

Fire deaths generally result from a combination of events in which fire originates at a source that ignites some type of fuel either in a part of the vehicle or on the roadbed after a collision. This is coupled with a simultaneous inability of passengers to leave the vehicle owing to a variety of circumstances. To prevent such fire-related deaths requires that the chain of events be broken at some stage in the process prior to the death of the occupants. An "event or consequence tree" is a useful tool in increasing fire safety since it makes it easier for the design engineer to understand the fire events and decide on the most economical method for preventing the occurrence of an undesirable process. An example of such a broken chain of events is prevention of occupant death by design changes leading to the fire-hardening of the fire wall of the engine compartment. This isolates the fire from the passenger compartment until the occupants can leave the vehicle. Conversely, any design changes that would either increase the intensity of the engine compartment fire or reduce the effectiveness of the fire wall in containing the fire could be expected to lead to an increase in the frequency of fire-related deaths. Similarly, the incidence of deaths from fires in the passenger compartment could be reduced either by design changes in the interior furnishings leading to a reduced frequency of fires in this area or changes in the vehicle doors leading to easier egress from the passenger compartment or both.

An NFPA report (1973) indicated that the Oregon Office of the State Fire Marshal, which has been keeping detailed records of passenger vehicle fires since 1971, reported that the most frequent ignition sources were backfires, electrical short circuits, hot mufflers and exhaust pipes, smoking materials, and incendiarism (in that order). Materials first ignited are gasoline and other flammable liquids, electrical insulation, and upholstery. Data such as these are valuable for indicating necessary design changes that can be made to improve the fire safety characteristics of automobiles.

9.2.2.1 Engine Compartment

Engine compartments are the source of 55 to 60 percent of all passenger car fires because both highly flammable fuel and ready ignition sources are present. A common source of fires in this area are leaky or ruptured fuel lines that spray flammable fuel on hot exhaust manifolds.

Inadequately protected electrical wiring passing through the engine compartment fire wall also is a common pathway for propagation of engine compartment fires to the passenger compartment. The fire-hardening of such electrical pathways is one important way in which fire safety of passenger vehicles could be improved at relatively low cost.

An analysis of the present and future applications for polymers indicates that a significant portion of the polymers used in passenger automobiles can be found in the engine compartment; even more of the metal components are expected to be replaced by various plastics, predominantly polypropylene, by 1980 (see Section 9.4). The substitution of polymeric materials for metal components under the hood, or in forming engine compartment boundaries should be carefully considered to prevent an increase in the fire load in this fire-susceptible area.

9.2.2.2 Passenger Compartment

As noted above, the passenger compartment has the second largest incidence of non-collision-related fires. This fact is the more significant since the compartment does not contain any source of flammable fuel. Fires in this area can be considered directly related to flammability of the interior furnishings and to the frequent presence of small ignition sources such as smoking materials. Since it is essentially impossible to control the ignition sources in this area, design changes leading to improved fire safety must be concerned with changes in the interior materials — in this case, polymeric materials since modern car interiors are finished almost completely with synthetic polymeric materials (see Section 9.4). Despite the high frequency of passenger compartment fires, the frequency of fire-related fatalities or injuries is probably low because of the easy egress that would be expected under normal circumstances once the vehicle is brought to a complete stop; however, there are few if any statistical data currently available to substantiate this conclusion.

9.2.2.3 Fuel Tank Area

A consideration of the frequency of fires in the fuel or storage compartment indicates that it has a negligible frequency in non-collision-related fires but it has second place in collision-related fires. These statistics are understandable if one considers that the fuel tank can contain between 20 and 25 gallons of highly flammable fuel that generally is not available for a fire until the tank is breached by some force, generally during a rear-end collision.

Since the fuel itself is so combustible and is present in such large amounts, the presence or absence of relatively small amounts of flammable polymers in this area will have little or no appreciable effect on fire safety once the fuel tank has been breached and the spilled fuel ignited. Bragman (1970), in a rather extensive test series on the susceptibility of automobile fuel systems to leakage in barrier crashes, has shown that a wide variety of fuel system failures could be induced by a strike on any part of the car. Design changes to improve the fire safety of this part of the vehicle must be predominantly concerned with hardening the fuel tank to prevent rupture by rear-end and other collisions. MVSS 301 was promulgated during 1968 in an attempt to reduce the frequency and severity of collisions leading to fuel fires. As indicated by the statistics summarized in Figure 1, the regulation has had little or no success in accomplishing this objective. More effective safety standards and concomitant design changes are required to improve fire safety or reduce fuel-fire-related fatalities and injuries.

Some design changes already have been suggested to improve the fire safety of the fuel tank area. Several of these changes are mechanical and have little relation to the fire safety of polymeric materials and are not discussed. Fabricating the fuel tank and fuel lines out of impact-resistance polymers, such as nylon or polyethylene may very well lead to a more impact-resistant fuel compartment and significant improvement in fire safety. A study to determine the effects of polymeric fuel tanks on the overall fire safety of the fuel compartment should be made before a large-scale conversion to the thermoplastic polymer tanks is implemented.

9.2.3 Passenger Car Fire Scenarios

9.2.3.1 Collision-Related Fuel Compartment Fire

Fire of this type usually originates from a rear end collision that ruptures or severely damages the fuel tank to such an extent that most or all of the fuel tank contents are spilled on the vehicle or in the vicinity of the crash. Ignition can occur from many sources, including electrical sparks from the other car. Sparks generated from the scraping of metal on metal is perhaps the most common ignition source. Whatever the combination of circumstances, the car's interior generally turns into an inferno in a matter of seconds. A series of actual rear end collision tests by the Insurance Institute of Highway Safety (1973) has recently been completed and in six of the full-scale tests, large fuel leaks resulted; one of the six resulted in spontaneous ignition leading to the almost complete burnout of both vehicles.

Collisions resulting in the rolling over of a vehicle also often lead to fuel tank rupture and ignition. Although less frequent than rear end collisions, such crashes have a considerably higher incidence of fire-related fatalities because the deformation of the roof that almost always results from such a roll-over often jams the doors and inhibits the escape of the occupants.

9.2.3.2 Engine Compartment Fires

As noted earlier, engine compartment fires are the most numerous of automobile fires, both collision- and non-collision-related. Although there has not been a detailed analysis of the causes of these fires, Siegel and Nakum (1970) have indicated that 90 percent of engine compartment fires were caused by malfunctioning of the carburetor. Frequently these fires are caused by the removal of the air cleaner or the loss of carburetor fuel chamber plugs.

9.2.3.3 Passenger Compartment Fires

Passenger compartment fires are the second most frequent type of fires and only seldom result from collisions. Smoking materials and defective wiring are the most frequent causes of fires in this area (Trisko 1975b). Fire-hardening of the interior furnishings would be the most effective way of reducing fires resulting from ignition by smoking materials.

An actual passenger compartment fire (Portland, Maine Press-Herald 1974) is recounted below:

A mother and four children had gone to the local dump to deposit trash and to hunt for an old bicycle from which to obtain a set of handle bars. While the mother and oldest son were searching, they looked up and discovered that their 1968 station wagon, containing the three youngest children, was afire. The mother sent the oldest son to a nearby house for assistance while she attempted to rescue the children. her efforts were frustrated by the heat and smoke. When assistance arrived a few minutes later, the fire was so hot that the windshield was melted and the three children were dead. The Fire Chief concluded that the fire started in the passenger compartment although the engine had also been burned. The rear exterior of the car and the gasoline tank were untouched. No apparent exterior ignition source could be found and the car was parked away from accumulated trash. There was no indication of the recent burning of rubbish near the vehicle. An autopsy by a local pathologist indicated that smoke inhalation was the cause of all three fatalities.

Another example of an actual passenger compartment fire occurred recently in

California (personal communication from P. Cooley, Highway Safety Research Institute, University of Michigan, Ann Arbor). A passenger vehicle, proceeding at cruising speed on a local freeway, was set afire when a defective windshield wiper was activated. The fire spread with such rapidity throughout the interior of the passenger compartment that one of the occupants in the rear seat was unable to escape by the time the vehicle was stopped; he died in the ensuing fire.

9.2.4 Fire Spread

There is no known data on the rate of fire spread in passenger car interiors. One can deduce conclusively, however, that materials conforming to MVSS 302, with a flame spread of 4 inches per minute in a horizontal configuration, will burn at a rate several times faster when placed in a vertical configuration such as is experienced on seat back and wall coverings. That the actual flame spread rate can be dangerously fast is indicated by the fact that all passengers in the second scenario in Section 9.2.3.3 were unable to exit safely from a vehicle fire that was initiated under ordinary conditions while the vehicle was proceeding at turnpike speed.

In addition, the use of less flammable materials in the passenger compartment would have the economic benefit of reducing the property loss in many small automobile fires by preventing the rapid flame spread currently observed and by allowing the economic repair of many vehicles now being scrapped after fire exposure.

9.3 Automobile Fire Tests and Standards

9.3.1 Introduction

Until a few years ago, there was no government safety regulations relating to private passenger automobiles and the fire safety characteristics of the finished vehicle was left to the discretion of the individual automobile manufacturer. Passage of the Highway Safety Act by Congress in 1070 charged the then newly formed National Highway Traffic Safety Administration (Department of Transportation) with the issuance of such safety standards as were deemed necessary to reduce the high incidence of death and injuries associated with highway accidents. During the ensuing years, several federal Motor Vehicle Safety Standards (MVSS) have been promulgated concerning various aspects of the motor vehicle safety problem. Of the several standards proclaimed, only two, MVSS 301 and MVSS 302, are concerned with the reduction of the high incidence of motor vehicle fires. Of these two, only MVSS 302 attempts to regulate the flammability of polymeric materials used in the construction of passenger cars, trucks, buses, and multipurpose passenger vehicles.

9.3.2 Motor Vehicle Safety Standard 302

MVSS 302, which became effective on September 1, 1972, was designed (according to the stated purpose of the standard) to reduce the yearly incidence of deaths and injuries to motor vehicle occupants caused by motor vehicle fires "especially those originating in the interior of the vehicle from sources such as matches and cigarettes."

The standard test procedure designated in this test is a modification of ASTM D1992 test for the measurement of the flammability of plastics. The test is conducted in a metal cabinet 15 inches long, 8 inches deep and 14 inches high. The cabinet is equipped with a glass observation window, a closable opening for inserting the specimen, a hole to accommodate tubing for the gas burner, and 10¾-inch ventilation holes in the base. Conditioned test specimens, 4 by 4 inches in size, are placed between two rectangular-shaped frames within the draft-free cabinet, in a horizontal position and ignited by a 1½-inch-high gas flame adjusted with the air inlet at the base of the burner closed. The flame is positioned at the unsupported end of the U-shaped frame so that the center of the burner tip is ¾ inch below the open edge of the specimen for a 15 second interval. The rate at which the flame front burns down the specimen for a distance of 11 inches, or wherever it stops prior to reaching a point 1½ inches from the clamped end of the specimen, is then calculated by the formula:

$$\text{Burning Rate (inches/minute)} = 60 \times \frac{\text{Distance burned (inches)}}{\text{Time (seconds)}}$$

Materials with burning rates of 4 inches per minute or less are considered to be satisfactory by this test.

As has already been noted (see Section 5.2.1.1), this test is extremely lenient and materials with 4 inch per minute burning rates in the horizontal position can be extremely flammable in many actual applications. This is especially true if the material is used in a vertical configuration as many materials are used in car interiors (e.g., for seat backs and sidewall liners). It is the opinion of the committee that the fire safety of the car interiors could be improved considerably or, conversely, that the flammability of interior materials could be reduced considerably by requiring the materials to pass the UL94 test conditions with a V-O rating. It is estimated that such a more stringent fire safety specification could be met by current state-of-the-art materials with only a modest increase in material cost — perhaps 10 to 30 percent on a weight basis. The PVC seat covers could be formulated to pass a UL94 V-O rating by the substitution of a phosphate plasticizer for the phthalate plasticizers currently in use. The difference in cost between the two types of plasticizers is 12 to 26 percent on a weight basis.

9.3.3 FAR 25.853, Flammability Test for Polymeric Components

The FAR 25.853 test procedure consists of exposing, in a vertical position, at least three specimens (each greater than 2 inches wide and 12 inches long cut from the actual material to be used in the vehicle and in the thickness to be employed) to a Bunsen burner flame 1½ inches in height. Exposure to the flame shall be for a period of 60 seconds for *Section-a* designated materials (ceiling panels, wall covering, partitions, etc.) and 12 seconds for *Section-b* designated materials (floor coverings, textiles, seat cushions, etc.). Section-a materials must be self-extinguishing by this test, with an average burn length not to exceed 6 inches and anaverage flame time not to exceed

15 seconds after removal of the flame. Drippings from the specimen, if any, may not continue to flame for more than an average of 3 seconds after falling. Section-b materials also must be judged self-extinguishing by this test, but the average burn length may not exceed 8 inches, the average flame time may not exceed 15 seconds, and any drippings may not continue to flame for more than an average of 15 seconds after falling.

9.3.4 Smoke and Toxicity

No test standards are currently available or even under serious consideration for application to passenger automobiles. Because of the small volume of the passenger compartment and the relatively easy egress from it, except under some crash conditions, it is doubtful that any such standards would significantly reduce the number of fatalities or injuries associated with passenger compartment fire even if such standards were set.

9.4 Polymeric Materials in Passenger Automobiles

9.4.1 Introduction

Although the increasing frequency of motor vehicle fires described in Section 9.1 is undoubtedly related to the increasing number of passenger vehicles, it also coincides with the increasing use of polymeric materials in automobile construction. The increasing use of plastics has been accelerated recently by the onset of the energy crisis. This crisis has accelerated a movement to lighten automobiles as a means for decreasing the amount of fuel being consumed. One way to decrease automobile weight is to increase use of plastic materials. These have already made extensive inroads in the automobile industry.

9.4.2 The Extent of the Usage of Polymeric Materials in Automobile Manufacture

The extent of the increasing plastics usage is indicated in Table 4.

Table 4. Plastics Usage in U.S. Automobiles (Avg. Lbs/Car).*

	1960	1973	Estimated 1980	Estimated 1985
Plastics	22	138	300	400

NOTE: Data from Ward's Automobile Yearbook 1975.
*Estimates are from the Ford Motor Company pertaining to its own use, but it is believed that the data is typical for other automobile manufacturers.

Despite the severalfold increase in plastics usage between 1960 and 1973, polymers account for only 4 to 5 percent by weight of the materials of construction of the 1973 automobile. The probable importance of this increased plastics usage on the increasing fire frequency in motor vehicles is unknown at this time.

The types and amounts of plastics anticipated for use in 1980 motor vehicles, together with their major applications, are summarized in Table 5.

Table 5. Projected Plastic Consumption in 1980 Cars[a]

Plastic	Estimated lb/car, Formulated Plastic	Major Applications
ABS	17	Grilles, instrument panels, lamp housings, interior paneling, wheel covers
Acrylic	3	Lenses, escutcheons
Acetal	2	Tank caps, misc. hardware, underhood components
Nylon	8	Interior hardware, fender extensions, underhood components, bezels, elec. sockets
Polyethylene	7	Coolant system bottles, gas tanks, misc. trim, wire insulation
Polypropylene	35	Interior panels and trim, ducts, fender liners, lamp housings, battery cases, fan shrouds, heater and a/c housings, seat frames
Polyester, thermoset	35	Front ends, fender extensions, heater and a/c housings, scoops, spoilers, some lid and door paneling, structural members
Polyester, thermoplastic	5	Valence panels, fender extensions, lamp housings, ignition system, interior hardware, misc. electrical
Phenolic	6	Ignition and electrical parts, brake and transmission parts, some pulleys
Urethane, flex foam and RIM	40	Seating and padding, flexible fascia and body parts, full-foam bumpers
Polyvinyl chloride	20	Upholstery, roofs, interior trim and panels, wire insulation
Other[b]	20	
	198	

[a]Excludes paint, tires, sealants, small elastomeric parts.
[b]Includes polycarbonate, cellulosics, polysulfone, styrenics, Noryl, alkyds.

As can be seen from this table, ABS, polypropylene, thermoset polyesters, polyurethane foams and solid elastomers as well as polyvinyl chloride comprise just less than 75 percent of the projected plastics usage. A considerable portion of this material will be going into car interiors for such applications as seat covers, foam cushion, instrument and trim panels, fascia, seat frames, and roof panels.

Although the data in Table 5 are reproduced from the commercial literature, their general accuracy has been confirmed by similar data obtained directly from at least one of the largest automobile manufacturers.

9.4.2.1 The Use of Polymeric Materials in the Engine Compartments of Passenger Automobiles

An analysis of the data presented in Table 5 indicates that the polymer that will be used most widely in the engine compartment (under-the-hood applications) is polypropylene (largely reinforced with glass or other mineral fibers or fillers). Thermoset phenolic resins, thermoplastic polyesters, nylon, and acetal are planned to be used to a much lesser extent and in much smaller quantities than is polypropylene in these applications. These latter polymers will find applications in such small parts as distributor caps, pulleys, miscellaneous electrical fittings, emission canister, and gasketing. As such, they cannot be expected to contribute significantly to the fire load in the engine compartment. Polypropylene, however, is to be used for fender liners, heating housings, battery cases, and fan shrouds requiring quantities measured in kilograms. The relatively high energy content of polypropylene (22 Kcal/g) and its ease of ignition could lead to a significant increase in the fire load. Additionally, polypropylene fender liners would not be expected to be much of a barrier to an engine fire because of relatively low melting point thus possibly exposing the tires as an additional fuel source during an early stage of an engine compartment fire.

9.4.2.2 The Use of Polymeric Materials in the Passenger Compartment

The passenger compartment of the modern automobile consumes more than half the total plastics used in the entire vehicle. About 21 pounds of flexible polyurethane foam is used in seat cushioning together with about 41 pounds of plasticized vinyl seat coverings. Approximately 16 pounds of ABS and 24 pounds of polypropylene are used in interior trim panels and head liners while about 30 pounds of SBR elastomers are used in floor mats. Other large-scale interior applications already in use or planned for inclusion in the 1977 or 1978 model years are side panels and load floors for station wagons and hatchback models, thermoplastic structural foam shells, ABS or SAN instrument panels and inner panels for doors. These applications add up to about 85 to 90 pounds of polymeric materials to be used in car interiors. The flammability of all these materials are little changed from the standard polymer by the need to conform to MVSS 302 burning rates and can be expected to considerably increase the fuel load in this compartment relative to interior construction in use only a few years ago.

Although all these materials are required to pass MVSS 302 with a horizontal burning rate of 4 inches per minute, most of them are used in a vertical configuration where the actual burning state would be expected to be several times that exhibited in the horizontal configuration. This fact, together with the high incidence of non-collision-related fires in this compartment, would indicate a high probability that the fire safety of this compartment may already be seriously compromised by the concentration of relatively high heat content flammable materials. Available statistics are not broken down sufficiently to allow an estimate of the annual property loss of fire injuries that are incurred from these non-collision-related passenger compartment fires.

9.4.2.3 The Use of Polymeric Materials in the Fuel or Trunk Compartment of Passenger Automobiles

Major applications for polymers in the fuel or rear section of the common passenger vehicle are valence panels, fender extensions, gas tanks, and fender liners. Although these components comprise a significant percentage of the total plastics content of the vehicle, the presence of highly flammable gasoline fuel in this area is probably the overriding fire hazard associated with the vehicle. Therefore, the relative fire hazard associated with polymer components is probably small compared to the hazard and fire load associated with the fuel in the tank. This conclusion is substantiated by the low incidence of non-collision-related fires noted in the statistical data summarized in Table 2.

9.4.3 Long Range Polymer Use in Automobiles

Experimental passenger vehicles have been designed and constructed during the past several years with bodies that are entirely, or almost entirely, composed of various polymer composites. Most of these vehicle generally are fabricated of foam sandwich constructions consisting of rigid polymer sheet on at least one side of the sandwich body and polyurethane, or some other structural foam, as the rigidizing filler. Although the advantages of low cost, low weight, fatigue resistance, and corrosion resistance of such materials have been well identified, the relative fire safety and flammability of such construction has not been determined. As forerunners of future vehicles, it will be necessary to carefully evaluate the fire safety and flammability characteristics of such construction before it is used in mass-produced vehicles. The committee can visualize some combinations of polymers and structural foams that could present a serious fire hazard to the life and safety of the occupants of such vehicles.

With the present lack of accurate information on the relative fire safety of all-plastic vehicles, it is considered to be essential that such information be developed in the near future if serious fire hazards in future automobile construction is to be prevented. Such safety evaluations should include the testing of full-scale vehicles as well as smaller-scale mock-ups to obtain realistic values of hazard.

9.5 Conclusions and Recommendations

Conclusion: The estimated annual 500 fatalities and $135 million property loss directly associated with passenger car fires are sufficiently serious to warrant considerable effort to improve the fire safety characteristics of automobiles.

Conclusion: Passenger compartment furnishings are almost 100 percent polymeric materials (predominantly flexible polyurethane, polyvinyl chloride, polypropylene and SBR elastomers).

Conclusion: Polymeric materials are major contributors to the high frequency of passenger compartment fires.

Conclusion: The increasing substitution of flammable polymers for metal in passenger car construction can be expected to increase the frequency and severity of passenger car fires unless adequate steps are taken to reduce the flammability of these materials by developing and enforcing more stringent flammability test standards.

Conclusion: The replacement of MVSS 302 with a more stringent flammability standard can be expected to significantly reduce the incidence and severity of passenger compartment fires at only moderate cost.

Recommendation: Require all polymeric materials used in automobile passenger compartments to pass FAR 25.853, Sections a and b (Sections b-1, b-2, and b-3 are not considered suitable), except for those components that can be shown to create no appreciable fire hazard.

Recommendation: Carefully evaluate the flammability characteristics of passenger vehicles with bodies constructed entirely or largely of synthetic polymers on full-scale mock-ups before such vehicles are mass produced for general consumption.

Recommendation: Undertake a study to determine the effect of plastic fuel tanks on overall fire safety.

CHAPTER 10

TRUCKS AND SPECIAL PURPOSE VEHICLES

10.1 Introduction

This chapter is concerned with the fire safety of several separate and diverse types of land transportation vehicles such as: highway property carriers (trucks), recreational vehicles, motorcycles and snowmobiles. It is intended to be primarily an overview because much of what needs to be written about the fire safety aspects of polymeric materials for these vehicles has already been included in preceding chapters. The problems are similar if not identical. Specifically, the materials available and used are the same, the tests are the same, smoke and toxicity considerations are the same or similar, and the fire paths through the vehicles are similar. Accordingly, the reader is referred to those chapters of this volume and other volumes of this series for detailed discussion of topics that are not covered in this chapter.

10.2 Highway Property Carriers (Trucks)

10.2.1 Background

Statistics for highway property carriers for the years 1970, 1971, and 1972 are presented in Table 1 and for the years 1973, 1974, and 1975 in Table 2. It can readily be seen that the fire related fatality rate and the property damage loss is a small percentage of the total problem.

Table 3 is a detailed analysis of truck fires for 1971. The Department of Transportation (1973) states: "The most frequent causes of fire continue to be hot tires, defective wiring, and collision impact." Table 4 indicates the numbers of fire accidents in which the vehicle was mechanically defective. At the time tires and electrical wires led the list for the fourth year in a row.

Although fires are a small percentage in the total statistics, they have not been overlooked by the National Highway Safety Administration (NHSA), which recently proposed a new federal motor vehicle standard that could help reduce the incidence of fires in motor vehicles. In advance of proposed rule making, NHSA called for information that could lead to regulation of electrical systems in motor vehicles. It stated: "Statistics indicate that between 500 and 750 deaths are caused every year by motor vehicle fires. Our Federal Motor Vehicle Safety Standard No. 301, which deals with fuel system integrity, regulates the spillage of fuel as a means of preventing vehicle fires in a crash, thus handling one major cause of vehicle fires. The establishment of performance requirements for electrical systems should combat the other major contributor to such fires."

Table 1. Highway Property Carriers. Accidents Involving Fire, 1970, 1971, 1972.

	Accidents	Fatalities	Injuries	Property
Moving Vehicles				$ Million
1970	544	125	277	6.441
1971	542	130	274	6.362
1972	495	113	226	7.100
Stopped				
1970	185	7	32	1.389
1971	169	10	37	2.400
1972	180	2	65	1.253
Totals (Fire Accident)				
1970	729	132	309	7.837
1971	71!	140	311	8.762
1972	675	115	291	8.353
Total Truck Accidents				
1970	50,387	1,961	23,174	105.245
1971	65,158	1,916	23,978	112.365
1972	62,229	2,087	25,866	129.499

Note: Data from Department of Transportation 1972, 1973, 1974.

Table 2. Highway Property Carriers.

PRIMARY EVENT - NON-COLLISION ACCIDENT

FIRE

	Accidents	Fatalities	Injuries	Property $ Million
1973	336	15	44	3.600
1974	287	3	34	3.626
1975	234	3	37	3.221

COLLISION AND NON COLLISION ACCIDENTS

TOTALS

	Accidents	Fatalities	Injuries	Property $ Million
1973	30,270	2,939	34,537	168.900
1974	32,014	2,722	31,575	210.860
1975	24,303	2,234	26,408	101.358

Note: Data from Department of Transportation 1972, 1973, 1974.

Table 3. Cause of Fires — Property Carriers.

Carrier's Vehicle

Cause of Fire	Moving	Stopped
Unknown or Not Reported	51	59
Other	7	10
Defective Wiring	55	34
Hot Tire	56	1
Hot Service Brakes	31	0
Friction	12	1
Battery Cable	5	2
Refrigeration Equipment (including Cab A/C)	0	2
Oil - Hot Parts	8	1
Bearings - Defective, Hot	17	0
Contact with Electric Wires	11	1
Collision Impact	145	6
Cargo Heater:		
Defective	8	
Defective	7	3
Too Close to Lading	8	6
Bombs or Arson	0	6
Leaking Fuel Tank:		
Hot Engine	1	0
Collison Impact	1	0
Hot Exhaust	1	0
Cigarettes, Fuses, Matches	0	1
Other	9	1
Smashed Fuel Tank:		
Collision Impact	47	0
Hot Exhaust	2	0
Sparks	5	0
Back Fire	1	0
Hot Engine	1	0
Leaking Fuel Line:		
Collision Impact	1	0
Sparks	1	0
Hot Engine	4	0
Hot Exhaust	3	0
Other	1	0

Table 4. Mechanical Defects — Property Carriers (by Major Assembly).

	Moving	Stopped
Axle-Steering	1	0
Axle-Drive	1	0
Axle-Non Drive	1	0
Body	15	3
Brake System	37	0
Transmission and Clutch	1	0
Electrical System	59	38
Engine	12	1
Suspension	4	0
Fuel System	18	3
Steering System	1	0
Drive Shaft	1	0
Tires	67	2
Instruments and Controls	1	0
Exhaust System	6	0
Wheel Bearing	17	1
Total	242	48

Note: Data from Department of Transportation 1973.

10.2.2 The Overall Problem

As with other land transportation systems, these vehicles are governed by MVSS 302 in addition to MVSS 301. If all other factors were to remain constant, the new performance requirements for electrical systems might prove to be adequate to reduce the incidents of fire. There are, however, driving forces that indicate "all other factors" probably will not remain constant.

The costs of transportation relate in part to the weight that must be moved, which of necessity includes the weight of the vehicle. The costs also relate in part to the periodic replacement costs of various parts of the vehicle. Historically, these costs have been based on metals; however, materials lighter and/or more durable than steel now are available. In some cases, these materials are being exploited; in others, they are being explored to bring down cost. There is also the national energy conservation drive aimed at reducing fuel consumption; this makes reduction in weight attractive as a national objective.

The technological and economic feasibility of a prototype graphite-fiber-reinforced plastic truck spring has been demonstrated. This item alone could save 400 pounds per truck, has a projected fatigue life of over 1 million miles; i.e., the life of the vehicle, and would enable the operator to reduce his costs by 200,000 ton-miles over the life of the truck. Similarly, a prototype drive shaft has been made out of graphite-fiber-reinforced plastic. Experimental models have been found to tolerate unbalance and to weigh half of a conventional part. General purpose insulated plastic truck bodies already are advertised for sale on the open market. Structural foams already are finding application as monolithic structural materials for commercial truck cabins. Other examples, such as the fiberglass-reinforced wheel covers and composite floor panels cited in Chapter 8, reflect the additional utilization of polymeric materials to reduce weight and costs.

Clearly, all-plastic trucks (except for engines) are now technologically possible. Furthermore, from the point of view of cost reduction resulting from fuel savings as a result of lighter weight or the increase in revenue rising from higher payloads, there is a socioeconomic benefit. If, as it now seems probable, greater fatigue-life materializes, there will be economic benefits in the replacement of parts. Similarly, the use of corrosion-free materials will reduce maintenance costs. These developments represent advantages that are available for trade-off against foreseeable disadvantages, one of which is susceptibility to fire.

The problem is to foresee and minimize the fire hazard resulting from the use of polymeric materials. With respect to the trucking industry, it is fortunate that the use of polymeric materials is in an early stage; accordingly, corrective actions should be easier to implement than in other industries where there is a vested interest in existing practices and materials.

10.2.3 Scenarios

Statistics are used by the National Highway Transportation Authority (NHTA) to develop and modify motor vehicle safety standards such as MVSS 301 and MVSS 302. (The problems in using statistics for developing standards have been discussed in Volume 4.) In brief, these statistics apply to an event happening at a given time in the prevailing environment. By environment is meant societal, technological, and physical circumstances and conditions. Changes in the environment could considerably change one's perception of the problem.

During the course of this study, statistics from which simplistic scenarios could be written were found to be freely available; however, scenarios illustrating the problems that might arise from using light-weight high-strength (energy efficient) materials were not available (trucks, however, are not unique in this regard). Some thought must be given to the development of predictive scenarios that will permit corrective action to be taken in the design phase rather than after a product is on the market or in use. Scenarios sensitive to political and economic developments or changes (e.g., shifts from an oil dependent economy to an oil depleted economy); to societal changes (e.g., shifts from *a smoking is required to a smoking is abhorred society);* and to technological changes (e.g., the development of hazard sensing and compensating devices to aid the operator) could be of great value to planners in both the public and private sectors. In other words, some effort should be directed towards planning for changes (scenarios) instead of all attention being focused on recorded events (statistics).

10.2.4 Materials

The polymeric materials that are used or will be used in trucks are discussed in this volume as well as in Volume 1. The generally unsatisfactory nature of most of these materials has already been discussed in Chapters 7, 8, and 9.

10.2.5 Test Methods

MVSS 302 controls the level of nonflammability that must be achieved before a material can be used. This standard was discussed in Chapter 5 and found to be inadequate. UMTA has issued more definitive guidelines for the selection of materials.

10.3 Recreational Vehicles (RVs)

10.3.1 Background

The term "recreational vehicle" is used here to mean the chassis and shell containing the contents of a typical trailer home or camper. The vehicle may be self-powered as in the mobile home or towed as in the trailer home or camper.

In one way this separation of frame from contents makes the problem simpler for the purposes of this section since the driving force toward plastics in the structural elements of the vehicle are the same as they are for the highway property carrier (i.e., lower weight means lower fuel consumption, nonmetallic parts mean less maintenance, greater fatigue strength means longer life and lower replacement costs).

A comprehensive and more cohesive picture of the fire safety aspects of recreational vehicles has recently been received, an excerpt from which follows:

> Fuel leakage from one or more of the involved vehicles was reported in about 7% of the accidents (071). Of the 69 reported occurrences of fuel leakage, 66 involved RV (44 Unit A vehicles, 20 Unit B vehicles, and 2 involving both Unit A and Unit B vehicles), and 3 involved other vehicles (072). In terms of vehicles, the incidence rate of fuel leakage for RVs was 4.7% (46 out of 981) for Unit A vehicles and 2.8% (22 out of 796) for Unit B vehicles. The overall incidence rate of RV was 3.7% (66 out of 1777) whereas, the incidence rate for other involved vehicles was only 0.5% (3 out of 564).

Seventy-five percent of the fuel leakage incidents involved gasoline, 20 percent involved leakage of LP gas, and the remainder involved leaks of other types of fuel. Ninety-three percent of the fuel leakage incidents were collision induced; only two cases involved non-collision (pre-cash) fuel leakage.

A recent study by the University of Utah Auto Crash Research Team examined post-crash factors including analyses of fuel leakage and vehicle fires in a sample of 13,000 traffic accidents involving over 23,000 vehicles. The incidence rate of fuel leakage was reported to be less than 2 percent in terms of vehicles, which is considerably lower than the rate observed for RV.

The high incidence rate of RV fuel leakage can be attributed to several factors such as the presence of auxiliary fuel systems for heating and cooking in most self-contained RV's, additional and/or large capacity fuel systems in many self-powered RV's, and the lack of crashworthiness of many RV's. Additionally, the current amended MVSS 301, which specifies tests for fuel system integrity, does not apply to multipurpose vehicles, trucks, and buses manufactured before September 1, 1976 or to auxiliary fuel systems in RV trailers and many self-contained self-powered RV's. Furthermore, vehicles with a gross vehicle weight rating of 6,000 to 10,000 pounds will not be required to meet the static rollover tests requirements included in MVSS 301.

Vehicle fires were reported in 28 (2.9 percent) of the RV accidents. Two of the fires started in another vehicle and spread to the RV after rear end collisions. Arson was reported in another RV utility trailer fire. Excluding these three cases, the incident rate of RV fires was 2.6 percent (25 out of 969 accidents), which is about eight times the rate observed in the aforementioned utah study. Of the RV fires, 12 involved trailers, 10 involved self-powered RV, and 3 were passenger car fires. In terms of vehicles, the incidence rate for trailer fires was 1.5 percent, and for self-powered RVs it was 3.6 percent.

A basic summary of the RV fires investigated is provided in Table 5. Note that all but two of the trailer fires were non-collision (pre-crash) fires whereas, Unit A vehicle fires were primarily collision induced. Only three of the collision-induced fires involved other (non-RV). vehicles. Over half of the RV fires were fuel-fed; 27 percent were reported as being electrical fires, primarily associated with battery sparking and/or shorting of the battery cables.

The extent of damage to individual vehicles shows that 14 of the 25 vehicles (56 percent) were "totaled" and 3 (12 percent) sustained moderate damage as a result of fire. Very few fire-induced occupant injuries were reported; however, property damage losses were very large in terms of vehicle costs and personal property losses.

In the discussion of fuel leakage presented above, it was noted that MVSS 301, Fuel System Integrity, did not apply to multipurpose vehicles, trucks, and buses included in the RV accident study. MVSS 302, Flammability of Interior Materials, effective for passenger cars, multipurpose vehicles, trucks, and buses manufactured after September 1, 1972, was applicable to a large portion of the vehicles in the bilevel RV accident sample. MVSS 302 specified burn resistance rates for interior materials in vehicles, but the test requirements are designed to reduce the frequency of fires starting from sources such as cigarettes and matches. Obviously, this standard totally neglects collision-induced circumstances such as fuel leakage and other sources of ignition. The risk of fires in an RV is much greater than for other vehicles due in part to the number and type of interior materials such as carpets, drapes, furniture, wood paneling, etc.; the type and location of fuel systems; and the auxiliary electrical systems provided in many RV.

As stated previously, the frequency of RV fires was found to be eight times the overall rate of vehicle fires reported in the Utah study. The circumstances of trailer fires indicate that improvements in electrical systems and upgrading of material flammability specifications are needed while self-powered RV need improvements in crashworthiness and fuel system integrity specifications. This conclusion should not be interpreted to mean that MVSS 301 and MVSS 302 are unsatisfactory in every detail.

Table 5. RV Fires (25 Cases).

Type of RV or Vehicle that Caught Fire	Type of Fire	Area of Fire Origin	Type of Accident	Extent of Fire Damage
I. Collision Induced				
Chopped Van	electrical	battery	ran-off-roadway	minor
Pickup Camper	gasoline	camper	fixed object	unknown
Van Conversion	gasoline	fuel tank	fixed object	total
Passenger Car	gasoline	unknown	rollover	total
Van Conversion	LP gas	living quarters	rear-end/rollover	total
Passenger Car	gasoline	fuel tank	sideswipe	total
Passenger Car	gasoline	engine	sideswipe	total
Van Conversion	electrical	battery	ran-off-roadway	minor
Pickup Camper	electrical	camper	rollover	minor
Travel Trailer	electrical	battery	rollover	minor
Travel Trailer	electrical	battery	sideswipe/rollover	unknown
II. Non-Collision				
Pickup Camper	gasoline	engine		total
Pickup Camper	unknown	engine		moderate
Motor Home	gasoline	engine		total
Motor Home	gasoline*	engine		total
Travel Trailer	LP gas	living quarters		minor
Boat Trailer	gasoline	boat engine		total
Travel Trailer	electrical	living quarters		moderate
Utility Trailer	unknown	interior		moderate
Travel Trailer	LP gas	living quarters		total
Camping Trailer	electrical	unknown		total
Travel Trailer	unknown	living quarters		unknown
Utility Trailer	unknown	interior		total
Mobile Home	other	wheel bearing		total
Utility Trailer	unknown	interior		total

* driver attempted to prime carburetor
Pass

10.3.2 Commentary

The societal needs for mobility and "temporary" housing, for affordable individual housing, for relaxation out doors and for the fulfillment of dreams of travel are all served at least in part by the recreational industry which includes the RV industry.

The interplay between the industrial need for growth and societal needs have created a line of products directed towards comfort and economy. Unfortunately, fire safety considerations have not played an important role in the development of these products. There may be some question as to whether or not the degree to which the problem exists at this time warrants a serious effort to regulate the industry to ensure that fire safety is taken into consideration. There is, however, little question that fire safety in such vehicles can be improved.

10.3 Scenarios

Statistics or data from which fire scenarios can be developed are available. A range of ignition sources (e.g., cigarettes, matches, electrical short circuits and collision impact) have been identified. Conventional materials including fuel, plastics, fabrics, upholstering, cushioning, and carpeting that can serve as fuel for a fire are available. The configurations into which these materials are likely to be fabricated also are well known as are the general habits and attitudes of the occupants. The driving forces in the community that lead to the purchase of recreational vehicles as life-style altenatives are identifiable. The political forces in the national economy that lead to the need for fuel economy and lighter weight structures also are identifiable.

Although the vehicle portion is currently small compared to the contents in terms of use of polymeric materials, the future can and should be predicted in order to minimize the contribution of the vehicle to fire problems in recreational vehicles.

10.3.4 Materials (See Section 10.2.4.)

10.3.5 Test Methods

These are adequately considered in the excerpt presented above. See also Chapters 7 and 8 for further discussion.

10.3.6 Other

The apparent role of fuel containers in the fire situation is not dissimilar to the problems encountered with controlling crash fires in helicopters (see Volume 6). This problem was solved through structural isolation of the fuel tank and the use of valves that sheared and sealed under impact. Defensive design techniques such as these should be considered for all fuel tanks in mobile homes and auxiliary fuel tanks in trailer homes and campers.

10.4 Motorcycles

Although the types and amounts of polymers used in motorcycle construction are growing, they still represent a very small potential contribution to the overall safety degradation and only limited fire risk. Polymers are used in hand grips, seats, windshields, tires, some fenders and splash guards, electrical insulation, and decorative panels (in some cases gas tanks are made of polymeric materials). Accessories (e.g., baggage racks, tote bags, travel cases) are increasingly being constructed of polymers. The contents of the accessory cases (clothes, bedding, etc.) are largely polymeric in nature.

Despite the growing polymeric fuel load, the conditions of motorcycle operation are not conducive to fire initiation by the polymer itself and the polymer does not readily transmit fire. When the motorcycle is at rest, the polymers are subject to a normal environment in which they can be ignited or support spread of fire. This condition, however, seldom presents a large risk to public safety, except as the fire may be spread to other vehicles or buildings.

By far, the most serious fire hazard to public health and safety from motorcycles is related to fuel spillage, leakage, or ejection, particularly after a skidding accident or collision. Until steps are taken to sharply reduce such hazards to the motorcycle operator and the general public, there appears to be little reason to expend effort on improving the fire safety characteristics of the polymers used in motorcycle construction.

10.5 Snowmobiles

Snowmobiles, like motorcycles, are being constructed with increasing amounts of polymers. Polymeric materials are used in windshields, front skis and track treads, snow and ice guards, seats, electrical insulation, and decorative panels. Accessories for support of the operator and passenger usually are made of polymeric materials. Most of the materials carried (except passengers) also are polymers.

From a fire standpoint, the snowmobile operates in a favorable environment (i.e., cold weather), and air movement discourages ignition and fire spread. Snow piled on a flaming component smothers and drenches the object. At rest, outdoors, conditions are only slightly less favorable (inside storage presents problems similar to other warehousing situations).

Fuel spillage and ejection by collision (usually with an inanimate object) present fire hazards many times more severe than those of the materials of construction. Thus, until better fuels and fuel containment are devised, there is little incentive to address the fire safety characteristics of the polymeric materials used in construction.

10.6 Special Purpose Vehicles

10.6.1 Nonpowered Vehicles

Most of the nonpowered vehicles, such as bicycles, have relatively small amounts of polymers and operate in relatively favorable environments from a fire standpoint. A major exception to this statement is the class of vehicles used for transport of the sick, injured, and aged (e.g., wheel chairs, stretcher carts, gurneys). Polymers are increasingly used in such constructions. Because the occupants often are unable to leave the vehicle, often drop ignition sources (cigarettes, matches, etc.) and because the vehicles sometimes operate in an oxygen-enriched atmosphere, the polymeric materials used must be carefully selected to provide reasonable fire safety. The committee is unaware of any major effort, national or local, to develop vehicles for the sick or injured with superior fire resistance characteristics. Such a program is needed.

10.7 Conclusions and Recommendations

Conclusion: Use of plastics and reinforced plastics in highway vehicles such as trucks, recreational vehicles, and motorcycles is increasing and will continue to increase for a variety of reasons (i.e., increased fatigue life reduces the need for replacement parts, reduced corrosion lowers maintenance costs, the lower density of plastics [compared to metals] permits reduction in the overall weight of the vehicle which, in turn, reduces fuel costs for the operator and fuel consumption for the nation).

Conclusion: MVSS 302 was developed from statistics for vehicles constructed primarily out of metals, and there is real doubt whether these statistics remain applicable during the ongoing shift from nonflammable metals to flammable plastics.

Conclusion: Fire scenarios coupled with qualitative fire dynamic considerations have not been applied in analyses of conditions as they will exist and have not been utilized in the prediction and prevention of future hazards.

Conclusion: A recurring problem appears to be that fuel storage facilities on vehicles contribute to ignition and the spread of a fire under impact of a collision.

Recommendations: Reexamine motor vehicle safety standards in light of the materials used now and expected to be used in the future in transport systems.

Recommendation: Reexamine polymeric materials for their fire safety aspects in the system in which they are currently being used or are intended to be used.

Recommendation: Verify fire scenarios through experiments and assign test methods and material property values on the basis of these experimental scenarios. For an unverified test method, assign tentative test methods and material property values on the basis of a worst case situation.

Recommendation: Examine the crashworthiness of all fuel systems involving polymeric materials with a view toward minimizing the fuel system as an ignition source, as an aid to flame spread, or as a source of fuel for the fire.

CHAPTER 11

SOCIETAL CONSIDERATIONS

11.1 Introduction

The fire safety problem in the United States, as elsewhere, has not been approached on a systems basis. Some efforts and expenditures have been made, but this generally has been done on a component basis or in response to a major fire calamity. Most efforts have been directed toward "fire drills" for escape, spot extinguishment of flames (e.g., sprinklers, CO_2 flooding systems), and application of "improved" materials to decrease the danger of ignition. The support of governmental and private citizens groups has been fragmented and applied without useful priorities based on risk analysis. Only recently has there been national awareness of the increasing severity of the fire problem and an approach made to an attack on a systems basis (National Fire Center, RANN support, etc.), but funds have been limited as has been progress.

Few designers, manufacturers, regulators, operators, or users of land transport systems have had any training in basic fire safety technology. Until recently there has been no place to obtain such training. Even today there are only a few colleges, technical schools, or other institutions where formal training in one or more aspects of fire safety can be obtained. Few engineering and technical training institutions require "fire safety" courses for matriculation.

There is no basic methodology to approach fire safety. Partial approaches, sometimes self-defeating, are found in construction codes, building specifications, governmental regulations and guidelines, and requirements. These, more often than not, were developed from analysis of a catastrophe or in response to a specific need.

The committee already has commented in some detail on these matters and has indicated a fire safety national need for:

1. Education and training.
2. Direction and coordination of efforts.
3. Financial support.
4. Methodology development.
5. Communication.

(For further discussion see Volumes 4 and 6.)

Under the circumstances related above it is not surprising that, using the available inadequate state-of-the-art data base, designers, manufacturers, regulators and operators are capable of providing land transport vehicles that have only limited fire safety performance. Such vehicles can, and should be, improved by a "systems approach" to the problem. The committee has suggested such an approach (elsewhere in this volume and in Volume 4) and also provided specific suggestions relating to materials, tests, toxicity, and end products (aircraft, land and water transport, buildings, etc.).

Government and private persons (and organizations) have performed well considering the circumstances discussed above and without exception have shown their concern, interest, and desire to improve the situation. Thus, there is no criticism or adversary relationship intended in any of the committee's comments that follow or in any portion of this volume.

The committee's "cross grain" look at the entire American society from a polymer fire safety viewpoint, together with its systems approach to solutions, has resulted in many observations and suggestions. All are presented in what is hoped will be accepted as a constructive manner; there is no intention to deprecate any person or organization.

11.2 Polymer Use and Abuse in Transport Systems

Polymers, particularly those that are made by man to his specifications, are finding increasing acceptance and use in land transport vehicles. Most of the major forces affecting decisions where there can be a choice of metals or polymers are pushing such decisions in the direction of polymers. These forces include advantages for polymer over metal as:

1. Reduced cost.
2. Reduced weight.
3. Reduced forming/manufacturing problems.
4. Reduced corrosion from wet or salt ambients.
5. Reduced maintenance requirements.
6. Reduced energy requirements in production.
7. Reduced energy consumption in operations.

The recent rapid increase in the cost of energy also has reinforced that driving force for the substitution of polymers for metals in many vehicle applications. As indicated in Figure 1, the energy requirements for the production of a unit of polymer are only one-tenth to one-half that for such metals as aluminum or magnesium while energy savings to from 10 to 80 percent are still possible relative to steel, depending upon the polymer chosen. This energy advantage of polymers over metals also extends into the energy consumption of the vehicle because the lower density of polymers reduces weight of the finished product with a resulting saving in operating energy.

As a result, automobiles, for example, now using 100 to 150 pounds of polymer per car will, in five years, use 500 to 800 pounds per car. Further increased usage is virtually certain.

Society is greatly benefitting from the continued increase in the use of polymers. Unfortunately, there are some disadvantages.

The polymers currently being used in transport units add considerably to the fire load. The replacement of 1 pound of metal by 1 pound of polymers adds 8,000 to 18,000 Btu. A considerable portion of the increase in polymer usage has been within the

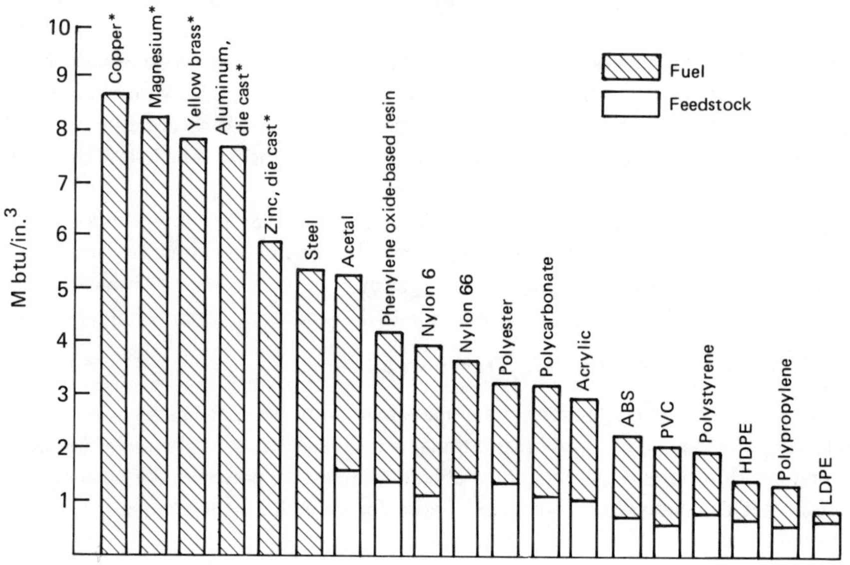

SOURCE: Du Pont
*Based on die cast industry estimates of secondary metals usage—20% magnesium, 46% aluminum, 5% zinc, 30% copper.

Figure 1. Energy requirements: production of plastics and metals.

passenger compartments or as elements of such structures. Since polymers are (relative to metal) easy to ignite and burn readily, land transport vehicles now are far more prone to fire damage and to creating situations in which passengers can receive serious burns. The fire hazards can be expected to increase as polymer usage increases unless corrective steps are taken.

Further, many polymers give off heavy smoke and toxic gases in a fire situation. Smoke and toxic gases seriously complicate and add tremendously to the probability of human injury in a fire. In many circumstances, the smoke and toxic gases are more serious hazards than the thermal effect of the fire.

Polymers have been substituted for metals in many places where such use is accompanied by benefits such as cost, reduced weight, etc., without full consideration of all of the functional uses of parts. For example, some urban transport buses have plastic wheel well covers; when fire occurs in the tire, brake or under the bus floor, it rapidly progresses through the polymer wheel well cover into the passenger compartment. A metal or more fire resistant cover would, in all probability, have denied access by the fire. Serious fires, involving fire transmission into the passenger compartment and resultant injury, have occurred in urban transit buses from just such circumstances.

Other cases of improper use of today's polymers are in bus and subway car floor structure, railroad car battery compartment structure, bulkheads between engine and passenger space, ventilation ducts (where they penetrate the floor or fire bulkheads), and thermal and acoustical insulation.

Interior designers are increasingly specifying polymers because of the exceptional decorative properties possible. Perhaps the misguided application of decorative polymers has led to the incorporation of unnecessary fire hazards into the interiors of transport vehicles. For example, carpeting material is being placed on vehicle compartment bulkheads in vertical orientations and in many cases on the ceilings; this sharply increases the fire hazard. The use of plush carpeting in place of rubber matting for subway car floors also has led to a substantial increase in the fire load and increased fire susceptibility.

The increase in fire hazard to the users of land transport vehicles and to the public in general has grown without general knowledge or appreciation and, apparently, without full evaluation by vehicle manufacturers. Such an evaluation by vehicle manufacturers and regulatory agencies is necessary and should include risk studies under varying circumstances. The scenario approach has been fully discussed earlier in this volume; it should be used extensively.

Few currently used composite polymeric materials are adequate where protection against full-scale fire is a functional requirement. The committee, at the risk of being considered parochial, believes that fire safety considerations are important to the public welfare and therefore recommends that materials in the key applications listed below be required to pass test ASTM E119 for 15 minutes duration until better tests, which consider operational needs, are designed and specified. Items to receive special tests described above are:

1. Structural floors (exception, steel jacketed, fire retarded plywood flooring).
2. Engine compartment bulkheads common to passenger compartment.
3. Wheel well covers.
4. Ventilation and other ducts, where they penetrate bulkheads (floor, etc.).
5. Bulkheads separating heavy electrical load devices (batteries, switches, etc.) from passenger compartments.

Other specific suggestions might result from fire scenario analysis of vehicle design.

11.3 Federal Support of Transport Systems

The energy crunch of 1973-1974 and its continuing adverse effects on single-person transportation have accelerated and given increased emphasis to the federal programs to support urban mass transit programs. Similarly, the national interest requires a viable railway system providing both passenger and freight service.

Large sums have been appropriated by the Congress to support these programs. Several large municipalities have developed subway or light rail transit systems; others have augmented rail and bus systems with new and more numerous vehicles. In many cases, substantial supporting funds have been provided by the federal government. In providing funds, the federal agencies have attempted to ensure public safety by setting mandatory requirements; however, the requirements set for fire safety have been found to be lacking on several counts:

1. The principal fire test method, MVSS 302, used as a criterion of safety, is inadequate for the purpose. As a result materials with poor fire characteristics have been allowed (polyurethane foam seat cushions, padding, etc.). At best MVSS 302 stipulates a slightly improved fire safety for the materials used (but has little system impact).
2. No limitations on total polymeric fire load, the position of the fire load, or its geometry have been established.
3. Fire dynamics in real fire situations have not been adequately studied.
4. Polymers have been improperly used (as previously discussed in Section 11.2).

The committee believes substantial improvements could be made if the federal agencies controlling funds set up adequate performance specifications in regulations for such public use items as subway vehicles, railroad cars, buses, etc. These specifications would become one condition for approving use of federal funds. A similar arrangement has been in use between the Navy, MARAD, and private shipbuilders in building commercial ships with defense features.

Transport vehicle performance specifications should include:

1. UMTA fire safety guidelines (as a minimum).
2. Limits on fire loads in each type of vehicle.
3. Operational requirements from a fire standpoint in sensitive applications.
4. Requirements for fire scenario development and analysis covering specific locations in mass transit cars and in each of the geographical areas transited (e.g., in a tunnel, on a bridge or elevated railway, on a highway).
5. Development of a fire hazard assessment for each new vehicle type and new transportation vehicle.

11.4 Uniform National Standards

Fire standards generally are set by local authorities in a particular city or state. Although many of these local entities follow standard codes set up by organizations such as the NFPA, there is much variation from community to community. Ground vehicles are built by manufacturers for national distribution and there is not so much difference in the operation of ground vehicles in any locality in the United States that different fire codes for materials or in fire safety standards are justified. It would seem to be in the interest of all of these communities and the public in general that minimum national fire codes should be promulgated for all transit systems, to include the vehicles and their environments, and that all systems be required to adhere to these codes.

11.5 Fire Extinguishment vs Fire Safe Design

Consideration is being given to the concept that it may not be necessary to require very low fire hazard characteristics of the polymeric materials if fire spread preventive measures are used. It is claimed that this will permit the use of less costly materials. Among such measures are fire extinguishing systems and air pressure differential procedures.

The dangers in this concept are that: (1) post-ignition control measures would permit a fire to occur and progress before extinguishment efforts were started and smoke and toxic gases would be released and fire damage would result; and (2) reliance on active systems presumes 100 percent reliability in the detection of the fire or smoke and the release of the extinguishant or that air compressors or fans will always work, but this is not the case as has been found in many fires.

For example, the U.S. exhibit at Montreal's Man and His World exhibition was built following the above concept with sheets of polymethyl methacrylate as the sheathing of the structure, on condition that the building be surrounded by water cannons. When ignition did take place, it was found that the water supply to the cannons had been turned off. The result was complete destruction of the building.

It is far better to concentrate on eliminating the risk of fire outbreak by the judicious use of materials and design as a passive means of restricting the spread of fire than to rely solely on extinguishing methods. This is particularly valid in transport vehicles where movement and distance invalidate many of the techniques for fire suppression used in stationary objects such as buildings.

11.6 Fire Safety-Biased Design Reviews

During many months of observation, study and analysis, the committee became aware of the interest in fire safety on the part of vehicle system regulators, operators, and manufacturers. Although knowledge of the fire characteristics of polymeric materials was limited, all supported improved fire safety through better polymeric materials, particularly if the added costs could be recovered.

All efforts in connection with polymeric materials to improve the fire safety of vehicles were on a *part* or *component* basis; other fire safety efforts were related to detection, extinguishment, or escape. There were *no efforts* to provide fire safety on an overall systems basis. So far as could be determined, no agency required a fire safety-biased design review of vehicles or system; although manufacturers wrote specifications, none conducted system design reviews to provide inherently safer vehicles or systems.

Current federal laws, and those of some states, require an environmental impact design review to reduce pollutants such as noise, particulates, and CO. Public health and safety are far more often and more seriously affected by fire in vehicles than by other environmental pollutants, but there is no requirement for a "fire impact" review.

While it is necessary to conduct large-scale tests to determine quantitatively the effects of burning various polymers used in vehicles, major reduction in fire hazards could be effected by conducting fire-biased design reviews of vehicles and their supporting components in transportation systems. Ideally, these reviews should be defined as a specification requirement in procurement documents and the results approved by the ordering activity (including UMTA when federal subsidy is involved) prior to the start of hardware manufacture. The scenario analysis technique should be an integral part of fire-biased design review.

If such review were performed, the nation could expect reduced vehicle fire hazard because of:

1. Improved polymeric materials.
2. Reduced fire load from polymeric materials.
3. Improved mechanical design of vehicle and systems.
4. Improved fire detection and extinguishment.
5. Improved system performance.

Similar reviews of existing systems could create priority listings for applications of resources to reduce fire hazard. Fire hazard assessment of existing systems should be undertaken immediately.

11.7 Economics of Fire Survivable Vehicles

Land transport vehicles currently being built are increasingly expensive; commuter cars, for example, approach $1 million per unit; buses are $65,000 or more per unit. The manner in which many are built virtually ensures the complete burnout of the interior from virtually any fire; worse, many are so constructed as to be non-refurbishable.

The committee is primarily concerned with hazard to the public and has already commented on the inadequacies of current fire standards and current construction methods from that standpoint.

It appears proper, despite being perhaps beyond the committee's basic charter, to invite attention to the heavy, unnecessary monetary and service losses that now are being incurred in vehicle fires. Since in many cases the financing of the vehicles comes from federal, state and local tax subsidies to operating transit systems, serious attention should be given to design and economic analysis of vehicles and other transportation system units to be more fire resistant and, thus less costly in the long term.

The suggestions made in Sections 11.2 and 11.3 would help substantially. In addition, consideration should be given to making mandatory the use of steel vehicle floors, side, and end walls; use of safety hydraulic fluids (silicones, etc., versus mineral oils), and, in transit systems, fire detection and extinguishment systems that can be operated remotely.

It has been observed that the urban transit cars manufactured by the General Electric Company at Erie, Pennsylvania (in 1976 in particular) contain virtually all the above suggestions; earlier vehicles of this type have survived fires that would have destroyed cars made by other manufacturers. If the suggestions of the committee relative to reduced fire load (i.e., through use of better polymers [e.g., neoprene versus urethane foam seats] and reduction of unnecessary decoration and extreme public comfort items) were adopted, these cars could be used as the fire safety standard for design evaluation of other vehicles. While these suggestions may result in higher unit first costs, current fire loss statistics indicate substantial long-term savings of public tax money.

11.8 Competence in Polymer Fire Safety Problems of Transit Operating Agencies

The committee has previously stated its concern that there is insufficient education and training effort concerning fire safety relating to polymeric materials (see Volumes 4 and 6). As a result of this situation and because fire safety is seldom identified as a specific item in specifications, transit operating authorities are not often blessed with outstanding competence in this field; this is not to say that there is a lack of *firefighting* competence. Since the state of the fire safety art is poorly defined at this time, the net result is that the level of fire safety competence relating to polymers in some transit authorities is no better than that of the average design engineer. Specification for new vehicles therefore are usually and unnecessarily deficient regarding polymer fire safety, particularly in connection with fire loads.

The efforts to improve fire safety by UMTA, FRA, and APTA are considered to be steps in the proper direction under difficult circumstances. These efforts need expansion with sizable additional resource allocations. Safety handbooks, training aids, and educational programs should be developed and offered to operating agencies to assist in preparation of better vehicle specifications. When these specifications require fire safe vehicles, they will be built.

11.9 Priorities and Understanding — What is Needed

There appears to be inadequate understanding of fire dynamics, fire probabilities, and public risks. There are inadequate assessments of consequences of certain kinds of actions in fire situations. Most fire safety actions of regulating and operating entities relate to fire drills and escape and specification of material characteristics. These are necessary but only a part of the totality of actions needed. As previously discussed, there also needs to be consideration of total fire load, position and geometry of part, fire dynamics (including effects of arson, vandalism, etc.), the number and types of persons occupying the vehicle, egress/escape routes, and use of fire statistics, probabilities, etc. These factors can be best understood and assessed by the use of the scenario technique (see Chapter 3). The scenario technique clearly defines the priorities and directions that fire-hardening must take in land transport vehicles.

It appears to the committee that the severity of the fire problem relating to polymers and therefore the priority for application of current resources and for additional effort in land transport vehicles should be (in descending order of importance) polymer fire safety problems relating to:

1. Subway vehicles and trains (normally carrying 100 to 1,000 persons with difficult egress and a confined space, e.g., a tunnel).
2. Elevated railway vehicles (normally carrying 100 to 1,000 persons with difficult egress).
3. Automated vehicles with no attendant (10 to 1,000 persons).
4. Light Rail (100 to 200 persons with reasonable egress).

5. Buses (40 to 80 persons with reasonable egress).
6. Automobiles (1 to 4 persons with good egress).
7. Trucks (1 to 2 persons with good egress).
8. Motorcycles, snowmobiles, etc. (with continuous, often involuntary egress).

11.10 Parting Thoughts

The use of polymers in land transport systems is growing by leaps and bounds. Although such growth appears to be in the national interest, the hazards of polymer usage have not been fully considered. A national need exists to accelerate the development of the methodology and the materials to solve this problem.

Appendix A

Proposed Guidelines for Flammability and Smoke Emission specifications — TSC-76-LFS-6 (Department of Transportation Memorandum, 6 July 1976).

The following proposed guidelines for flammability and smoke emission specifications are issued for application to combustible materials used in transit systems. The guidelines are revised periodically to reflect the certification of better standards and improved materials. Comments are solicited.

This comment supersedes TSC-75-LFS-5.

Scope — These specifications relate to all combustible materials used in a transit system, and include seats, seat cushions, upholstery, flooring, carpeting, wall and ceiling panels, plastic glazing, lighting diffusers, thermal and acoustical insulation, electrical insulation, elastomers and ducting.

1.1 Seat cushions and thermal and acoustical insulation shall be capable of passing the ASTM E 162-67 Radiant Panel Test with a flame propagation index (1) not exceeding 25. Additional provisions are as follows: (1) there shall be no flaming, running or dripping, (b) wire mesh screening shall be used (as per section 4.9.2 of ASTM E 162), (c) a 6-inch-long pilot flame shall be used (burner tip situated 1¼" beyond the frame to prevent extinguishment), (d) aluminum foil shall be used to wrap around the back and sides of the specimen.

The fire-resistant properties of the materials shall be demonstrated to be permanent by washing according to Federal Test Method 191 b, Method 5830.

1.2 Wall and ceiling panels, windscreens, seat frames, seat shrouds, partitions and ducting shall be capable of passing the ASTM E 162-67 Radiant Panel Test with a flame propagation index (I) not exceeding 35, with the added provision that there shall be no flaming dripping.

1.3 Upholstery materials shall be tested by FAA Regulation 25.853 vertical test, Appendix F(b), with the following modifications:

a) the average flame time after removal of the flame source may not exceed 10 seconds,
b) burn length shall not exceed 6 inches,
c) flaming dripping shall not be allowed,
d) fabrics that must be machine-washed or dry-cleaned must meet the requirements of 1.3 a), b), and c), after leaching according to Federal Test Method 191 b Method 5830, or after dry-cleaning according to AATCC[1]-86-1968. Fabrics that cannot be machine-washed or dry-cleaned must be so labeled and pass 1.3 a), b), and c) after being cleaned as recommended by the manufacturer.

---

[1]AATCC = American Association of Textile Chemists and Colorists.

1.4 Carpeting shall be tested with its padding, if the latter is to be used, and shall be capable of passing the NBS Flooring Radiant Panel Test, NBSIR-74-495, with a minimum critical radiant flux of 0.6 watts/cm².

1.5 Plastic windows and lighting diffusers shall be capable of passing the ASTM E 162-67 Radiant Panel Test with a flame propagation index (I) not exceeding 100.

1.6 Flooring shall be capable of withstanding the requirements of ASTM E 119 when exposed for 15 minutes up to 1400°F (760°C) on its underside.

1.7 Elastomers shall be capable of passing the requirements of ASTM C542-71A, with the added requirement that there be no flaming dripping.

1.8 Electrical insulation.

a) Wires for lighting, auxiliary circuits, speakers, public address, intercom system and the like shall be tested according to IPCEA-NEMA[2] S-19-81, paragraph 6.19.6 or Underwriters Laboratory Standard 62. The FR-1 restriction shall be applied to this test.

   Note: There is no standard test method for assuring circuit integrity of this type of wire during and after exposure to flame. However, it is required that an insulating char or residue remain on the specimen wires in order to maintain continuity of service.

b) High-voltage cable shall be tested according to the IEEE Standard 383-1974. A further provision of this test is that circuit integrity shall continue for five minutes after the start of the test.

2.0 Smoke Emission

Scope — This specification relates to all combustible materials as listed in 1.0 with exceptions as noted.

2.1 All materials shall be tested for smoke emission in accordance with the National Fire Protection Association Standard No. 258, "Smoke Generated by Solid Materials" (1974). The optical density, D, in both flaming and non-flaming modes, determined in accordance with the test, shall have the following limits:

a) For upholstery, air ducting, thermal insulation, and insulation covering, the D may not exceed 100 within 4 minutes after the start of the test.

b) For all other materials with the exception of cushioning, electrical insulation and carpeting, the D may not exceed 100 within 90 seconds after the start of the test, and may not exceed 200 within 4 minutes after the start of the test.

   Note: Test procedures for electrical insulation will be published as soon as such procedures have been finalized. In the interim, known heavy smoking insulation such as PVC and chlorinated, sulfonated polyethylene must be avoided.

3.0 Toxic Gas Emission

At the present time, there are no acceptable toxicity standards that can be applied to the types of materials listed above. It is hoped that such standards will soon become available, if only as preliminary standards.

---

[2]IPCEA = Insulated Power Cable Engineers Association, NEMA = National Electrical Manufacturers Association.

REFERENCES

Abu-Isa, I.A., J. Pol. Sci. Pt. A-1, 10, 881 (1972).

America Burning, The report of the National Commission on Fire Protection and Control, Mar. 1973 (Chap. 12)

Anon, Modern Plastics, 48 (January 1976).

Anon, Plastics World 41, November 17, 1975.

"An Appraisal of Halogenated Fire Extinguishing Agents," Proceedings of a Symposium Held at the National Academy of Sciences, April 11-12, 1972 (AD 753 218) National Technical Information Service, Springfield, Virginia, 22163.

Backus, J.K., and Gemeinhardt, P.G., in: Frisch and Saunders (1973), Rigid Urethane Foams, Vol. 1, Part II, pp. 451-524.

Ball, G.L., "Toxic Gas Evolution from Burning Plastics," Polymer Preprints 14 (2), 986 (1973).

Barechi, C.J., Paper, "Ignition of Bus Seats," May 1976.

Bateman, R., ed., "The Chemistry and Physics of Rubber-like Substances," John Wiley, New York (1963).

Benisek, L., International Dyer and Textile Printer, 414-19, April 7, 1972.

Benisek, L., Melliand Textilever., 8(6), 931 (1972a).

Benisek, L., Textileveredlung, 8(6), 318 (1973).

Billmeyer, F.W., Jr., "Textbook of Polymer Sciences," 2nd ed., Wiley Interscience, New York (1971).

Birky, M.M.; Coats, A.E.; Aldeson, S.E.; Brown, J.E.; Pabo, M.; Pitt, B., "Measurement and Observations of the Toxicological Hazards of Fire in a Metro Rail Interior Mock Up," NBSIR 75-966, February, 1976.

Blair, N.D., Witschard, G., and Hindersinn, R.R., J. Paint Techn., 44, 75-82 (1972).

Bragman, A.F., Jr., "Impact Intrusion Characteristics of Fuel Systems," Cornel Aeronautical Laboratory, Inc., Final Report, April 1970.

Brake, F.P., "Potential Tank Automotive Applications for Organic Materials," Army Materials and Mechanics Research Center, Watertown, Massachusetts, 02172, 31 March, 1976.

Braun, E., Report of Fire Test on an AM General Metro Bus, NBSIR 75-718, Center for Fire Research, Institute for Applied Technology, NBS, Washington, D.C., 20234, June 1975.

Braun, E., "A Fire Hazard Evaluation of Interior of WMATA Metrorail Cars" NBSIR 75-971, Center for Fire Research, Institute for Applied Technology, NBS, Washington, D.C., 20234, December 1976.

Breden, L. and Meisters M., NBSIR 76-1030, February 1976.

Brydson, J.A., "Plastics Materials," 3rd ed., Van Nostrand, New York (1975).

Burge, S.J. and Tipper, C.F.H., Combustion and Flame, 13, 495 (1969).

Canterino, P.J., in: EPST, 6, 432 (1967).

City of Los Angeles California Fire Department Report, "Origin of Fires in Passenger Automobiles," 1974.

Committee on Fire Safety Aspects of Polymeric Materials, Vol. 1 — Materials, State of the Art; Vol. 2 — Test Methods, Specifications and Standards; Vol. 3 — Combustion Toxicology of Polymers; Vol. 4 — Fire Dynamics and Fire Scenarios. Technomic Publishing Co., Westport, CT., 1977-78.

Clark, C.C., and Krawczyk, A., U.S. Pat 3,365,420 (1968).

Clark, C.C., Krawczyk, A., Reid, G.C., and Lind, E.V., Paint and Varnish Prod., 56-59, (April 1967).

Compton, F.A., Canadian Patent (to Dow Corning Co.) (1967).

Conley, R.T., and Gaudiana, R.A., "Thermal and Oxidative Degradation of Polyamides, Polyesters, Polyethers, and Related Polymers" in Thermal Stability of Polymers, Vol. 1, p. 347, ed. R.T. Conley, M. Dekker, N.Y. (1970).

Conley, R.T., and Malloy, R., "Vinyl and Vinylidene Polymers" in: Thermal Stability of Polymers, Vol. 1, p. 223, ed. R.T. Conley, M. Dekker, N.Y. (1970).

Conley, R.T., and Quinn, D.F., "Retardation of Combustion of Phenolic, Urea-Formaldehyde, Epoxy, and Related Resins Systems" in: Lewis, Atlas, and Pearce, (1975).

Conventional School Bus Design Objectives, School Bus Manufacturers Institute, Jan. 1973.

Cooley, P., Fire in Motor Vehicle Accidents, a special report of the Highway Safety Research Institute, April 1974.

Cooley, P., Highway Safety Research Institute Reports, 5 (1), September 1974.

Cornish, H.H. and Abar, E.L., Toxicity of Pyrolysis Products of Vinyl Plastics, 1969.

Creitz, E.C., J. of Res., Nat. Bur. Stds. 65A (4), 389 (1961).

Creitz, E.C., J. of Res., Nat. Bur. Stds., 74A (4), 521 (1970).

Cullis, C.F., "Combustion of Polyolefins," Oxidation and Combustion Rev., 5, 83-133 (1971).

Dalzell, D.A., and Nulph, R.J., Society of Plastics Engineers, 28th Annual Technical Conf., Vol. XVI, p. 215, New York (May 1970).

Discussion at Wayne Corporation (school buses) and Flexible Div., Rohr Corporation (commuter buses).

Dow Corning Co., British Patent 1,161,052 (1969).

Drake, G.L., Jr., "Fire Resistant Textiles," in *Encyclopedia of Chemical Technology,* 2nd Ed., Vol. 9, pp. 300-15, Wiley-Interscience, N.Y., 1966.

Drake, G.L., Jr., "Fire Resistant Textiles," in *Encyclopedia of Chemical Technology,* 2nd Ed., Supplementary Vol., pp. 944-64, Wiley-Interscience, N.Y., 1971.

DuPont Magazine, Airway Instruction, pp. 14-17, Mar-Apr. 1976.

Einhorn, I.N., *Fire Retardance of Polymeric Materials, Polymer Conference Series,* University of Utah, June 14, 1970.

Einhorn, I.N, "Fire Retardance of Polymeric Materials," J. Macromol. Sc. — Revs. Polymer Technol. D1(2), 113-184 (1971).

Essenhigh, R.H. and Howard, J.B., Ind. & Eng. Chem. 58, 15 (1966).

Fabris, H.J., and Sommer, J.G., "Flame Retardation of Natural and Synthetic Rubbers," in: Kuryla and Papa (1973).

Fenimore, C.P. and Jones, G.W., Combustion and Flame 10, 295 (1966).

Fenimore, C.P. and Martin, F.J., Combustion and Flame 10, 135 (1966).

Foy, G.F., "Engineering Plastics and Their Commercial Developments" in: Advances in Chemistry, American Chemical Society Monographs, No. 96 (1969).

Friedmay, M. et al., *Text. Res. J.,* 43, p. 212, 1973.

Freidman, R., "A Survey of Knowledge About Idealized Fire Spread Over Surfaces," *Fire Research Abstracts and Reviews* 10, 1-8 (1968).

Fistrom, R.M. Chem. & Eng. News, 150 (Oct. 14, 1963).

Fryburg, G., *Review of Literature Pertinent to Fire Extinguishing Agents and Basic Mechanisms Involved in Their Action,* NACA T.N. 2102 (1950).

Galloway, Anolick, Stewart, McSweeney and Johnson, Paper, "Flammability of Neoprene Cushioning Foam and the Neoprene Interliner System in Seating," May 1976.

Gmitter, G.T., Fabris, H.J., and Maxey, E.M., in: Chapter 3, Vol. 1, Part I, pp. 109-226. Flexible Polyurethane Foam, Frisch and Saunders (1973).

Gouinlock, E.V., Porter, J.F. and Hindersinn, R.R., J. Fire & Flammability 2, 206 (1971).

Gross, D., Loftus, J.J., Lee, T.G. and Gray, V.E., "Smoke and Gases Produced by Burning Aircraft Interior Materials," Building Science, Series 18, National Bureau of Standards (1969).

Hagnauer, G.L., and Schneider, N.S. J. Polymer Sci. A-2, 10,699 (1972).

Halpern, C., J. Res. Nat. Bur. Stds. 67A(1), 71 (1963).

Hastie, J.W., Combustion and Flame 21, 49 (1973).

Hecker, K.C., Rubber World, 159(3), 59 (1968).

Hilado, C.J., "Flammability Handbook for Plastics," Technomic Publ. Co., Inc., Westport, Ct. (1969).

Hindersinn, R.R., "Fire Retardancy," in Encyclopedia of Polymer Science and Technology, Wiley-Interscience, N.Y., 1977.

Hindersinn, R.R. and Wagner, G.M., Encyclopedia of Polymer Science and Technology, Volume 7; Interscience Publishers, Division of John Wiley & Sons (1967),

REFERENCES

Hindersinn, R.R., and Wagner, G.M., "Fire Retardancy," in *Encyclopedia of Polymer Science and Technology,* Wiley-Interscience, N.Y., 1967.

Hofman, W., "Nitrile Rubber," Rubber Chem. Tech. 36, 1-252, (1963).

Holmes, C.A., Flammability Tests of Selected Wood Products Under Motor Vehicle Safety Standards, 1973.

Hooker Chemical Corp., (a) Preliminary Data Sheet No. 349, "Dechlorane 604" (August 1970). (b) Techn. Bull. Dechlorane Plus, 602 and 604 as Fire Retardant Additives for Elastomers (1970).

Hooker Service Bulletin -2100; Durez Division, Hooker Chemicals & Plastics Corp.

Howarth, J.T., Lindstrom, R.S., Sheth, S.G., and Sidman, K.R., "Flame Retardant Additives," Plastic World, p. 64-67 (March 1973).

Insurance Institute for Highway Safety, Washington, D.C., Status Report 8 (11), May 29, 1973.

Insurance Institute for High Safety, Washington, D.C., Status Report 10 (3), February 5, 1975.

Johnson, P.R., 3rd International Cellulose Plastics Conference, SPI Preprints, Montreal, Canada; Sept. 27, 1972.

Karstedt, B.D., U.S. Patent 3,539,530 (1970).

Kasem, M.A., and Rouette, H.K., J. Fire and Flammability, 3, 316-29, Technomic Publ. Co., Inc., Westport, Ct. 06880 (1972).

Kennedy, J.P., and Tornquist, E.G.M., eds., "Polymer Chemistry of Synthetic Elastomers," Interscience, New York (1968).

Kimmerle, "Aspects and Methodology for the Evalulation of Toxicological Parameters During Fire Exposure,": 1974.

Krajewski, D.J., Hooker Chemical & Plastics Corp., Unpublished work.

Krekeler, K. and Klimke, P.M., *Kunstoffe,* 1965.

Lask, G. and Wagner, H.G., *8th Symposium on Combustion,* Williams and Wilkins, Baltimore, Maryland (1962), p. 432.

Laur, T.L., and Guy, L.G., Rubber Age, 102 (12), 63 (1970).

Lauriente, M. and Wiggins, J.H., "A National Program for Fire Safety in Transportation," 4th Paper Presented at the 4th Intersociety Conference on Transportation, Los Angeles, California, p. 3, July 20, 1976.

Learmonth, G.S. and Nesbitt, A., British Polymer J. 4, 317 (1972).

Learmonth, G.S., Nesbitt, A. and Thwaite, D.G., Brit. Pol. J. 1, 149 (1969).

Learmonth, G.S. and Thwaite, D.G., British Polymer J. 1, 154 (1969).

Learmonth, G.S. and Thwaite, D.G., British Polymer, J. 2, 249 (1970).

Ibid., British Polymer J. 2, 104 (1970).

Letter, Department of Safety, City of Cincinnati to Urban Mass Transit Authority, re bus seats, March 5, 1976.

Lewen, M., Atlas, S., and Pearce, E.M., "Flame Retardant Polymeric Materials," Plenum, N.Y., 1975.

Lindemann, R.R., "Fire Retardation of Polystyrene and Related Thermoplastics," in: Kuryla and Papa (1973).

Little, R.W., *Flameproofing Textile Fabrics,* ACS Monograph Series No. 104, Reinhold Publishing Corp., New York, N.Y. (1947).

Lyons, J.W., "The Chemistry and Uses of Fire Retardants," Wiley-Interscience, New York, 1970.

MacKay, D.J., Fire J. 64, 52 (March 1970).

Madorsky, S.L., *Thermal Degradation of Organic Polymers,* Interscience Publishers, (1964), p. 160.

Magee, R.S., and McAlvey, R.F., III, "The Mechanism of Fire Spread," *Journal of Fire and Flammability 2, 271-96 (1971).*

Marciniak, H.W., Hooker Chemicals & Plastics Corp., Unpublished work.

Mark, V., U.S. Patent 3,940,366 to General Electric, 1975.

Modern Plastics, (Plastics in Transportation), June 1976.

Morton, M., ed., "Rubber Technology," Reinhold, New York (1973).

Nametz, R.C., Ind. Eng. Chem., 59, 99 (1967).

National Fire Protection Association, "Transportation Fire Hazards," Staff Study SPP-19, Boston, Massachusetts, p. 25-29, April 1973.

National Transportation Safety Board, Safety Recommendation, H-76-7, issued March 14, 1976: H-75-39 issued Dec. 23, 1975; H-75-12 and H-75-13 issued June 9, 1975.

National Transportation Safety Board, Safety Recommendation H76-7 of March 14, 1976, "A Report on A Washington Metro Bus Fire on Shirley Highway, November 20, 1975; Washington, D.C.

Nouvertne, W., U.S. Patent 3,775,367 to Farbenfabriken Boyer, 1973.

Palmer, H.B. and Seery, D.J., Combustion and Flame 4, 213 (1960).

Pariser, R., McEvoy, J.J., and Johnson, P.R., "Elastomers and Flammability," Tutorial Lecture for NAS Ad Hoc Committee on Fire Safety Aspects of Polymeric Materials, Elastomer Chemical Department, E.I. DuPont de Nemours and Co. (Inc.), (1974).

Pearce, E.M., Shalaby, S.W., and Barker, R.H., "Retardation of Combustion of Polyamides," in: Flame Retardance of Polymeric Materials, Lewin, Atlas, and Pearce, (1975).

Pepe, A.E., German Patent 1,936,345 (1970).

Pigott, K.A., "Polyurethanes," in: EPST 2,506 (1969).

Pitts, J.J., J. Fire and Flammability, 3, 51 (1972).

Plastic World, Aug. 1976, (Market Growth for Polypropylene).

Polosar Inc., Data Sheet/600-R "Polysar SS 1905" (1970).

Portland, Maine Press Herald, August 17, 1974.

Potential Tank Automotive Applications for Organic Materials, P.F. Brake, Army Materials and Mechanics Research Center, Mar. 1976.

Proceedings, Conference on the Development of Fire Resistant Aircraft Passenger Seats, NASA Technical Memorandum, NASA TM X-73, 144 Aug. 1976.

Reid, N.J. and Heighway-Bury, E.G., J. of Soc. Dyers and Colorists, 74, 823 (1958).

Report on the Committee on Fire Safety Aspects of Polymeric Materials, Vol. 1 — Materials, State of the Art; Vol. 2 — Test Methods, Specifications and Standards; Vol. 3 — Combustion Toxicology of Polymers; Vol. 4 — Fire Dynamics and Fire Scenarios.

Rhys, J.A., Chemistry & Industry 187, (1969).

Rhys, J.A. and Cleavert, R.F., Plastics and Rubber Weekly 20 (Nov. 13, 1970).

Roberts, A.H., Haigh, D.H. Rathsack, R.J., J. Appl. Polymer Sci. 8, 363 (1964).

Rosser, W.A., Wise, H. and Miller, J., *Seventh International Symposium on Combustion,* Butterworth Scientific Publications, London, p. 175, 1959.

Rosser, W.A., Inami, S.H. and Wise, H., Combustion and Flame 10(3), 287 (1966).

Ryan, N.W., Polymer Conference Series Preprints, University of Utah, June (1970).

Saunders, J.H., and Frisch, K.C. "Polyurethanes," Part I, Chemistry, Interscience, New York (1962); Part II, Technology (1964).

Schmidt, W.G., Trans. J. Plastics Institute, 33, 247 (1965).

Siegel, A.W. and Nakum, A.M., "Vehicle Post Collision Considerations," International Automobile Safety Compendium, 1970.

Siegfried, K.J., in: EPST 7,432 (1967).

Simmons, R.F. and Wolfhard, H.F., Trans. Faraday Soc. 52, 53 (1956).

Specification (AMSI B73-5).

Stern, H.J., "Rubber: Natural and Synthetic," Palmerton, N.Y., (1967).

St. Louis Post-Dispatch — Thursday, May 27, 1976.

Stuctz, D.E., Symposium on Flammability Characteristics of Polymeric Materials, University of Utah, June 21 (1971).

Sumi, K., and Tsuchiya, Y., J. Fire and Flammability, 4, 15 (1973), Technomic Publ. Co., Inc., Westort, Ct. 06880.

Sunshine, N.B., "Flame Retardance of Phenolic Resins and Urea- and Melamine-Formaldehyde Resins," in: Kuryla and Papa (1973).

REFERENCES

Tesoro, G.C., and Meiser, C.H., Jr., Textile Res. J., 40, 430-36 (1970).

Tesoro, G.C., Report to the National Bureau of Standards, NTIS — Com 73-11265 (March 1973).

Texas-U.S. Chemical Co., Belgian Patent 729,226 (1964); British Patent 1,010,145 (1964).

Trexler, H.E., "The Formulation of Nonburning Elastomer Compounds," Rubber Chem. Tech. 46, 1114-1125 (1973).

Trisko, E.M., "Results of the 1973 National Survey of Motor Vehicle Fires," Insurance Institute for Highway Safety, Washington, D.C., p. 8, January 1975.

Trisko, E.M., Fire Journal 19, March 1975.

Vandersall, H.L., "Intumescent Coating Systems, Their Development and Chemistry," J. Fire and Flammability 2, 97-140 (1971).

Ward's Automotive Yearbook, 37th ed., p. 36, 1975.

Wayne Corporation Promotional Literature, "The Lifeguard Bus," 1976.

Weiner, S.A., American Chemical Society, Div. of Organic Coatings and Plastics Chemistry Preprints 34 (1), 55, (1974).

Whitby, G.S., Davis, C.C., and Dunbrook, R.F., eds., "Synthetic Rubbers," John Wiley, New York (1954).

Widmer, G., in: "Amino Resins," EPST, 2, 1 (1965).

Winspear, G.G., ed., "The Vanderbilt Rubber Handbook," R. T. Vanderbilt Company, New York (1972).